TRANSACTIONS

OF THE

AMERICAN PHILOSOPHICAL SOCIETY

HELD AT PHILADELPHIA

FOR PROMOTING USEFUL KNOWLEDGE

VOLUME XXII—NEW SERIES

PHILADELPHIA
PUBLISHED BY THE SOCIETY
1925

LANCASTER PRESS, INC.
LANCASTER, PA.

Contents of Volume XXII

PLATE IV.

FIG. 23.   Portion of larval pellet of *Pseudomyrma gracilis* (var. nov.), showing fragments of insects and pollen.   × 58.

FIG. 24.   Portion of larval pellet of *Pseudomyrma filiformis*, showing fragments of insects and spores. × 78.

FIG. 25.   Portion of larval pellet of *Pseudomyrma championi*, showing fragments of insects, spores and pollen.   × 78.

FIG. 26.   Portion of larval pellet of *Pseudomyrma flavidula*, showing fragments of insects, hyphæ and spores.   × 330.

FIG. 27.   Portion of larval pellet of *Pseudomyrma gracilis* var. *mexicana*, showing fragments of insects and hyphæ.   × 58.

FIG. 28.   Portion of larval pellet of *Pseudomyrma gracilis*, showing fragments of insect.   × 78.

FIG. 29.   Portion of larval pellet of *Tetraponera allaborans*, showing fragments of insects and pollen. × 78.

FIG. 30.   Portion of larval pellet of *Pseudomyrma gracilis* var. *mexicana*, showing fragments of insect. × 78.

FIG. 31.   Portion of larval pellet of *Pseudomyrma gracilis* var. *dimidiata*, showing fragments of insects. × 58.

PLATE V.

FIG. 32.   Portion of larval pellet of *Pseudomyrma gracilis* var. *dimidiata*, showing spores of different types among others those of *Ravenelia distans* A. & H.   × 330.

FIG. 33.   Portion of larval pellet of *Pseudomyrma rufomedia*, which is composed entirely of pollen and spores.   × 330.

FIG. 34.   Same as Fig. 32.   × 58.

FIG. 35.   Portion of larval pellet of *Pseudomyrma* species, from Patulul, Guatemala, showing dirt and other detritus.   × 58.

FIG. 36.   Portion of larval pellet of *Pseudomyrma gracilis* var. *mexicana*, which is composed entirely of spores and hairs.   × 330.

FIG. 37.   Portion of larval pellet of *Pachysima aethiops*, showing plant hairs, spores and dirt.   × 58.

FIG. 38.   Portion of larval pellet of *Pseudomyrma belti*, showing pollen.   × 58.

FIG. 39.   Portion of larval pellet of *Pseudomyrma decipiens*, which is composed almost entirely of spores.   × 330.

FIG. 40.   Portion of larval pellet of *Pseudomyrma sericea* var. *fortis*, showing hyphæ and fragments of medullary tissue from myrmecodomatia of *Triplaris macombii*.   × 330.

ARTICLE 5.

A REVIEW OF THE DORADIDÆ, A FAMILY OF SOUTH AMERICAN NEMATOGNATHI, OR CATFISHES.[1]

(Plates I–XXVII.)

By CARL H. EIGENMANN.

INTRODUCTION.

The present is one of a series of monographs dealing with the families and sub-families of the freshwater fishes of South America. Preceding monographs are: (1) A revision of the South American Nematognathi, or Catfishes. *Occasional Papers California Academy Sciences*, 1890, Vol. I. In coöperation with Rosa S. Eigenmann; (2) The Gymnotid eels of tropical America. Max Mapes Ellis, *Mem. Carnegie Mus.*, 1913, Vol. VI; (3) The Cheirodontinæ, a subfamily of minute characid fishes of South America. *Mem. Carnegie Mus.*, 1915, Vol. 7; (4) The American Characidæ. *Mem. Mus. Comp. Zoöl.*, 1917–1925, Parts 1, 2, and 3 issued so far; (5) The Pygidiidæ, a family of South American Catfishes. *Mem. Carnegie Mus.*, 1918, Vol. 7.

The present paper is based principally on the collections in the Museum of Indiana University.

The material was gathered during (*a*) my visit to British Guiana, 1908; (*b*) to Colombia, 1912; (*c*) the trip of Dr. W. R. Allen to the Peruvian Amazon, and (*d*) the trip of Mr. N. E. Pearson with the Mulford Expedition to Bolivia. In part it was received in exchange from other museums. In 1888 I examined the collection of Doradids in the Museum of Comparative Zoölogy (*Occasional Papers Cal. Acad. Sci.*, I, 1890), and more recently those of the Carnegie Museum. I am indebted to the Carnegie Museum, the Museum of Comparative Zoölogy, the Museum of the Phila-delphia Academy of Natural Sciences and the Tropical Research Station under the direction of Dr. William Beebe for the privilege of examining some of the Doradids in their collections. Dr. Thomas Barbour was kind enough to reëxamine *Hoplodoras uranoscopus* described by Mrs. Eigenmann and myself thirty seven years ago. Dr. Alipio Ribeiro generously had Sr. E. Cruzlima make drawings of the types of *Mega-lodoras libertatis* and *Anadoras insculptus*. Unless otherwise indicated the photo-graphs were made by the author, mostly based on anatomical preparations made by him. The drawings were made by Eugene N. Fischer (Plate XII), Dr. Clarence

[1] Contribution from the Zoölogical Laboratory of Indiana University, No. 198.

Kennedy (various details in the text figures), and particularly by W. S. Atkinson of Stanford University.

The numerous anatomical preparations enabled me to resurrect and re-define many of Bleeker's abandoned genera and to describe a number of other genera as new. The examination of the air-bladders demonstrated such an unexpected, amazing and riotous divergence that it is regrettable that not all of the species could be examined. So many species are known only from the types that probably many other forms are undiscovered, which makes rather elaborate if not complete illustrations justifiable and desirable.

POSITION OF THE DORADIDÆ AMONG OTHER SOUTH AMERICAN FISHES.

South America is very rich in freshwater fishes. As in North America, Europe, Africa and India, the Ostariophysi are dominant in South America. There are series of families of Heterognaths or Characins, another series of peculiar eels, the Gymnonoti and another series of Nematognaths or Catfishes. Of the fifteen families of catfishes found in South America, one family extends north to Central America, one is found in warm seas, the rest are all peculiar to South America.

Class   PISCES
Superorder or Series   OSTARIOPHYSI (Plectospondili)
Orders NEMATOGNATHI—HETEROGNATHI—GYMNONOTI.

South American Families of Nematognathi and their distribution:

DIPLOMYSTIDÆ (Temperate South America),
ARIIDÆ (Warm seas and rivers),
AUCHENIPTERIDÆ (Tropical South America),
DORADIDÆ (Tropical South America),
AGENEIOSIDÆ (Tropical South America),
HELOGENIDÆ (Tropical South America),
HYPOPHTHALMIDÆ (Tropical South America),
PIMELODIDÆ (Central America and Tropical South America),
BUNOCEPHALIDÆ (Tropical South America),
ASPREDINIDÆ (Tropical South America, in part marine),
PYGIDIIDÆ (South America),
NEMATOGENYIDÆ (Chile),
CETOPSIDÆ (Lowlands of Tropical South America),
ASTROBLEPIDÆ (Andes from Panama to Titicaca),
CALLICHTHYIDÆ (Tropical South America, chiefly lowlands),
LORICARIIDÆ (Tropical South America, lowlands and mountains).

THE NATURE OF THE DORADIDÆ.

Diagnostics.—The Doradidæ are a family of catfishes, Nematognaths, peculiar to tropical South America. They are distinguished from all other Nematognaths by the presence of a series of plates along the sides, each with a strong, median, backward-directed spine, sometimes supplemented by smaller spines on the surface of the plate. They are furthermore distinguished from all but the Auchenipteridæ by the exaggerated development of the parapophyses of the fourth vertebra, associated with

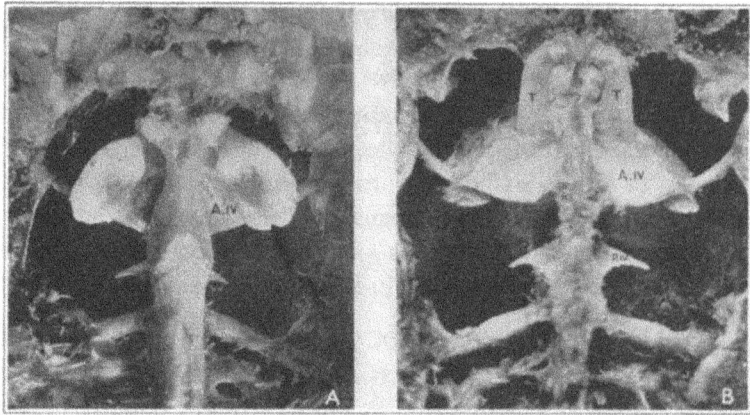

Fig. 1. The fourth vertebra of *Pterodoras granulosus*. *A*, From below, showing the anterior (*A*. IV) and posterior (*P*.) lateral processes. The anterior process with the oval disks which fit into pockets of the air-bladder. *B*, From above, the body of the vertebra partly cut away to show the Weberian bones. *T* = tripus or malleus. Compare with Fig. 11.

the air-bladder and the Weberian apparatus. The anterior lateral process of the fourth vertebra arises as a thin, firm, flexible plate and ends in a large, circular, disk-like or conical plate so closely attached in an indentation of the air-bladder that it frequently parts company with the basal part and remains attached to the air-bladder, when the latter is removed. The disk is of very different texture from that of the flexible, compact, basal part to which it is attached. The Doradidæ further agree with the Auchenipteridæ in the solidly united bones of the skull, nuchal shields and dorsal plate. The nasal openings are remote. The gill-membranes united with the isthmus.

The species reach an extreme length of a meter. Most of them are much smaller.

Structural radiation.—There are two groups of species arranged under "*a*" and "*aa*" in the key to the genera. In the broad-breasted species, "*a*," in which the head is more depressed, the premaxillary is normal, provided with bands of teeth, the mental barbels simple. In the narrow-breasted species the head is compressed, rather than depressed, the snout narrow, sometimes prolonged, conical. In these the premaxillary is subcircular, with very few or no teeth. Frequently the barbels are fimbriated and can be folded forward and used as a screen over the small mouth.

In the broad-mouthed species the anterior nostril is very near the lip, in the conical, prolonged-snouted ones the anterior nostril is always a considerable distance from the lip. There are no nasal barbels in any of the species.

The species best fitting the North American conception of a catfish is *Pterodoras granulosus* (Plate VIII). It has a broad ethmoid and broad premaxillaries, which bear wide bands of teeth. While in general appearance this species seems most conventionalized, the structure of its air-bladder indicates that it is far from primitive. Judging by the air-bladder and by the distribution, *Centrochir crocodili* and *Franciscodoras marmoratus* are the oldest and most nearly like the original Doradids. A very close rival to them is the widely distributed *Platydoras costatus* (Plate IX).

Fins.—The dorsal is usually composed of one spine and six rays. In the species of *Nemadoras bachi, Agamyxis pectinifrons,* and in *Acanthodoras,* it is I, 5, even I, 4 in some specimens of *A. cataphractus* and in *castaneoventris,* which is probably a synonym of the *cataphractus*. It is situated immediately behind the nuchal shield. The dorsal spine may be slender or large and heavy, serrate on both edges, serrate only in front, or without serræ. The spine, if without serræ, has longitudinal grooves and ridges. The anal is short, with from 10 to 16 rays. The adipose fin is present in all but *Physopyxis.* Usually it is well defined and short. In the larger, heavily armored species it is usually not well defined, prolonged forward as a heavy and scarcely flexible keel, extending from the adipose toward the dorsal. The caudal may be rounded, truncate, emarginate or deeply forked. The ventrals are rounded, without marked peculiarities. The pectoral spine is in some species notably heavy and large, in others long and slender. It is always serrate on both margins, sometimes also on the dorsal surface.

Bones of the head.—The bones of the upper surface of the head are granular or striated, covered with very thin skin. They are united by sutures into a solid shield or plate. The sutures can frequently be seen only with a lens. In the numerous figures the sutures were traced and marked under a binocular dissecting microscope. The bones are uniformly numbered in the different figures. The bones of the top of the skull united by sutures are (2) the ethmoid, (3) lateral ethmoid, (5) frontals, (6) sphenotic, (7) pterotic, (8) supraoccipital, (9) epiotic and three elements (*X, Y, Z*) of the dorsal scutes, which are expanded ends of interneural spines.

The skull varies from nearly flat in *Acanthodoras* to sharply roofshaped in *Trachydoras* (Plates II and IV). The suprascapula (10) is ankylosed to the pterotic

(7) and epiotic (9). It entirely excludes the epiotic from the margin of the skull in the broad-snouted species and in *Pseudodoras*. In all the species in which the epiotic (9) is connected with the first lateral scute by a process (9a) it forms part of the margin of the shield and the supraclavicle (10) is but narrowly in contact with the epiotic (9).

In the broad-headed species the fontanel is a small, oval opening between the anterior end of the frontals (Plates II to IV), sometimes extending into the ethmoid. In extreme cases the fontanel is reduced to sub-circular (Plate II, Fig. 3) and may be entirely occluded. In the narrow-headed species the fontanel is more elongate. It

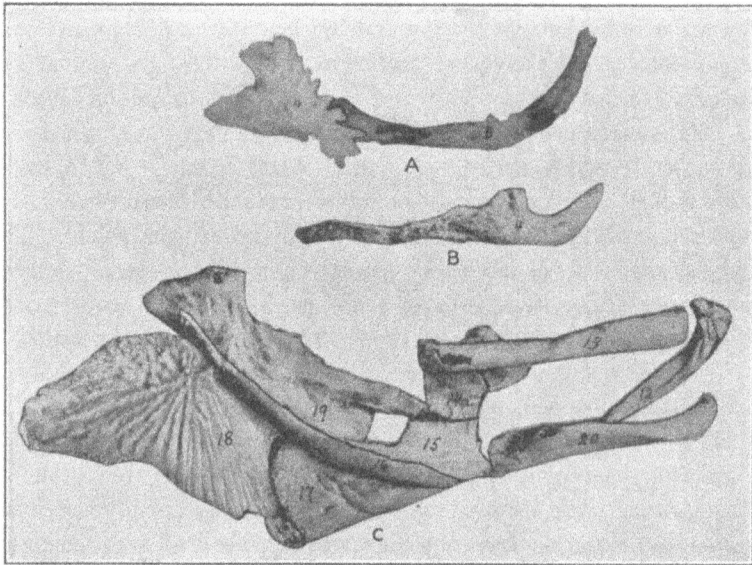

FIG. 2. *A*, Preorbital and suborbitals of *Astrodoras asterifrons*. *B*, Preorbital and suborbitals of *Pseudodoras niger*. *C*, Membrane bones of the right side of *Pseudodoras niger*. Notations: 4, Preorbital; *A* and *B*, suborbitals; 12, maxillary; 13, palatine; 14, mesopterygoid; 14a = metapterogoid; 15, quadrate; 16, preopercle; 17, interopercle; 18, opercle; 19, hyomandibular.

moid and into the occipital, with a narrow bridge in the posterior third of the frontal. reaches its maximum in *Leptodoras* (Plate II, Fig. 2) where it extends both into the eth-

A very slender rod, the nasal, is attached by ligament to the margin of the ethmoid, between the ethmoid and the nasal cavity. It is marked *D* in the plates, but is mostly obscure.

There is but little variation of the same bone in the different species. The greatest variation is found in the epiotic and in the region in front of the eye, in the preorbital bone (4) and the suborbital chain. The preorbital bone (4) is always more or less fan-shaped. The pointed anterior end of the bone may be in contact with the maxillary and premaxillary. It may not reach so far. In a few species, *Pseudo-*

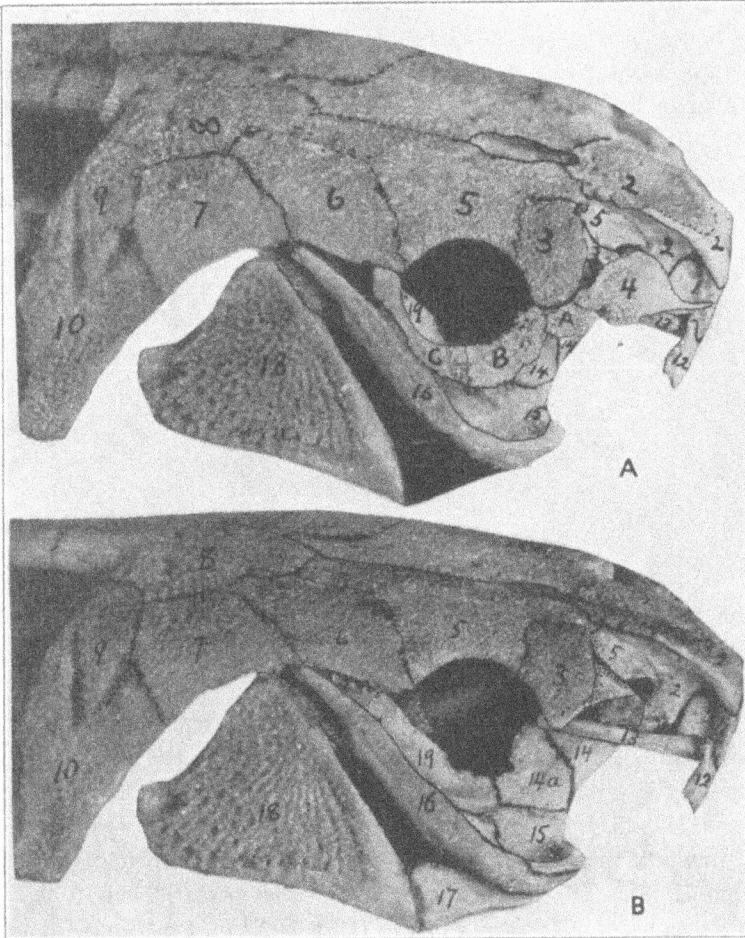

Fig. 3. *Platydoras costatus*. *A*, With the preorbital and suborbitals. *B*, With the preorbital and suborbitals removed.
1, premaxillary; 2, ethmoid; 3, lateral ethmoid; 4, preorbital; 5, frontal; 6, sphenotic; 7, pterotic; 8, supraoccipital; 9, epiotic; 10, supraclavicle; 12, maxillary; 13, palatine; 14, mesopterygoid; 15, quadrate; 16, preopercle; 18, opercle.

doras niger, it is joined to the lateral shoulder of the ethmoid, the ectethmoid. It is procumbent in the larger species, its expanded edge smooth (*Megalodoras*, Plate VI, Figs. 3 and 4) or scalloped and covered with skin. In other species it is serrate or finely pectinate (*Agamyxis*, Text-fig. 14, *Astrodoras*, Plate III), the pectinate edge being raised to form a crest in front of the posterior nostril. The suborbital varies with the size of the eye. In some species, those with minute eyes, the first suborbital lies under the preorbital and does not form part of the orbital margin. In other cases all three suborbital bones form the orbital margin and are of about uniform width. In some cases they are feeble and lie imbedded in the skin; in others they are of unequal size and one or all of them thick, granular.

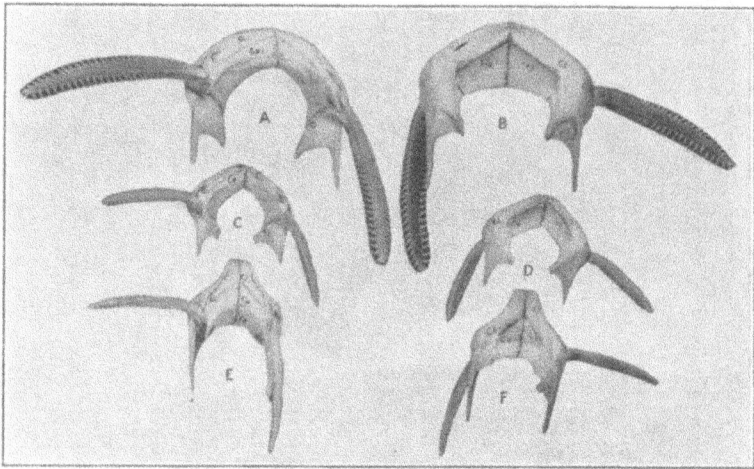

Fig. 4. Clavicle, coracoid and pectoral spine in *Megalodoras irwini*: *A*, from below; *B*, from above; *Pterodoras granulosus*: *C*, from below; *D*, from above; *Doras micropœus*: *E*, from below; *F*, from above.

The opercles may be covered with smooth skin, but usually the opercle (18 in the figures) is at least partly granular or striate, and sometimes the interopercle and preopercle are notably granular (*Trachydoras*, Plate XVIII).

The palatine (13) is rod-shaped, in contact with the maxillary in front (12) lying above the mesopterogoid (14), which connects the metapterogoid (14*a*) with the lateral ethmoid (3). The size of the palatine varies with the length of the snout, reaching its maximum in *Doras micropœus* (Plate XXIV), *Leptodoras* (Plate XXIV) and *Pseudodoras* (Plate XVII).

The shoulder girdle.—The clavicle and coracoid are always united with their fellows of the opposite side by an interlocking suture into a solid plate, provided with a process extending from the coracoid backward below the pectoral, and another process usually longer than the coracoid from the clavicle backward above the pectoral, the so-called humeral process.

Rarely the coracoid process is exaggerated and extends as far as the humeral process (*Trachydoras paraguayensis, cl.* and *co.*, Plate XVIII), or even much further (*Physopyxis lyra,* Plate 16). The surface of the humeral process may be granulate or the granules may be enlarged into spines along a series, particularly toward the tip. In the more exaggerated forms the spiny humeral process and the pectoral spine form a scissors-like apparatus (Plate XXV, Fig. 3).

The upward projecting flanges of the coracoid and the inward flanges of the clavicle are very different in different species. In the narrow-breasted genera they are extensive and form a dividing wall between the gill-cavity and the parts behind.

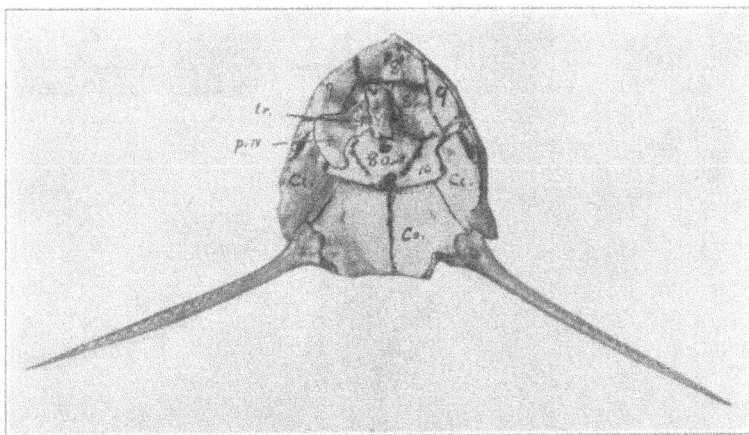

Fig. 5. Cross section of *Trachydoras atripes* (I. U. M. 16006) just behind the bony diaphragm formed by the flanges of the coracoid, *co.;* clavicle, *cl.,* supraoccipital 8, exoccipital 8*b*, basioccipital 8*a*; epiotic 9, supraclavicle 10; the broken vertebra *v*, the process of the fourth vertebra *X*.

In these species the effect is heightened by the expansion of the basioccipital into a thin plate extending downward. The partition or diaphragm formed by the flanges of the clavicle, coracoid and basi- and exoccipital reaches its maximum in the species of *Trachycorystes* (Text-fig. 5). The opening is just sufficient for the passage of the alimentary canal, being much smaller than the pupil. In *Trachydoras paraguayensis* the downward plate of the basioccipital and supraclavicle actually overlap, passing behind the processes of the clavicle and coracoid. In this species the opening is not visible looked at from straight behind; the alimentary canal is a siphon. The partition is least developed in *Megalodoras* and *Lithodoras*.

In most of the species the lower surfaces of the coracoid-clavicle are covered with the erector muscles of the pectoral spine and these with skin (*cl.* and *co.* in the ventral views on many of the plates); in a series *Anadoras grypus, weddelii* (Plate XV), *Ambly-*

doras monitor (Text-fig. 15), *A. hancockii* (Plate XIII) and *Physopyxis lyra* (Fig. 1, Plate XVI), there is a decreasing space for muscles and an increasing granulation of the bones, covered with very thin, scarcely perceptible skin.

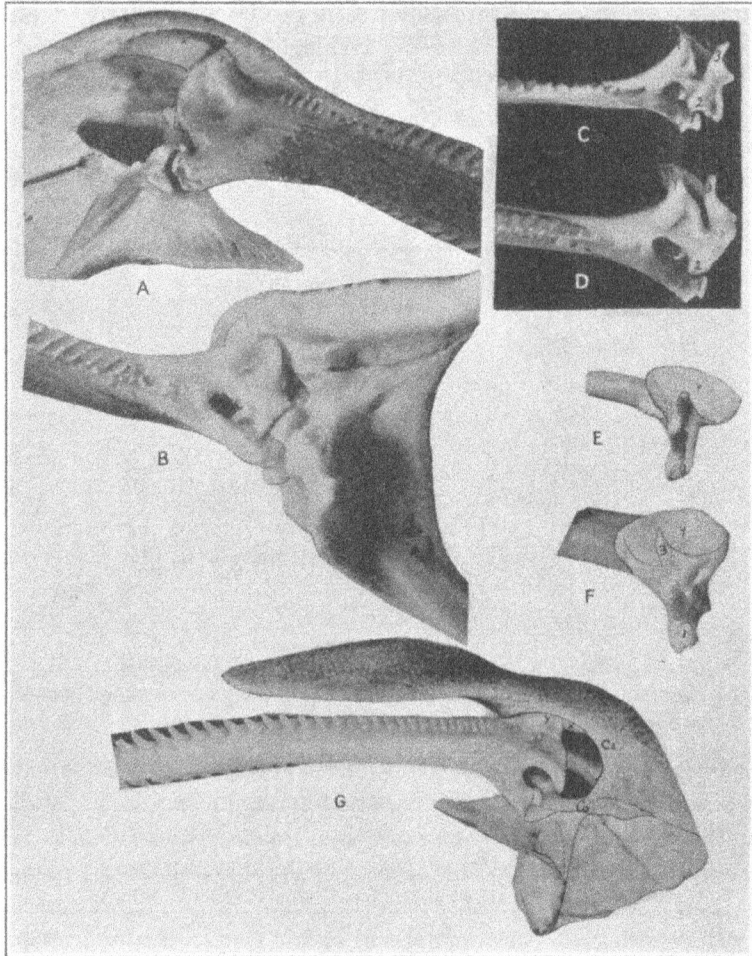

FIG. 6, *A, B, D*. Pectoral lock of *Megalodoras irwini. A*, Pectoral spine depressed; *B*, the same erect; *D* and *F*, the basal end of the pectoral spine, *D* from the axillary edge, *F* looking directly at the base of the spine. Part of the clavicle cut away to the left of the spine in *A*, to show the groove in which the outer flange of the spine moves. *C* and *E*, the basal end of the pectoral spine of *Pseudodoras niger*, to correspond with *D* and *F*. *G*, the right pectoral spine of *Platydoras costatus*.

Lock of the pectoral spine.—A most striking feature of the Doradidæ is the elaborateness of the pectoral spine and everything associated with it, its muscles, and the adaptation of the bones for their attachment, the coracoid and clavicle.

As in other Nematognaths the pectoral spine can be erected and locked at any stage of erection. Though in type the lock is not different from that in some other Nematognaths it may not be out of order to redescribe it.

The left pectoral joint and lock can best be imagined by using the right hand and arm as a model. The arm would represent the spine (Text-fig. 6), the thumb (1) and pointer finger (2) would represent two processes engaging between them a portion of the coracoid, the tip of the thumb turning in a socket in the lower outer surface of coracoid as a fixed point, the tip of the forefinger moving in a short groove of the inner surface of the coracoid.

The middle finger is grooved along its entire lower surface, fitting over a ridge of the coracoid, the end of the middle finger (3) impinging upon the clavicle.

The ring and little fingers (4 or y) are greatly expanded sidewise into a curved blade projecting around the dorsal surface of the spine, moving in a deep, *curved* groove in the expanded, laterally-projecting portion of the clavicle. In erection and depression the spine revolves on two axes. In the main movement, the erection and depression, the spine moves about the axis provided by the ridge over which the grooved lower surface of the spine fits. The second axis extends from the thumb to the tip of the spine and movement about it results in the tortion of the spine. The thumb (1) moves, twists in the socket, the pointer finger moves in a short, shallow groove in the coracoid, the expanded flange (4) moves in a much longer, deeper, curved groove in the clavicle. In order to depress, the spine must not simply revolve on the transverse (middle finger) axis; it must also undergo a slight tortion about the longitudinal axis of the spine with the end of the thumb as the pivot. The moment the twisting in the longitudinal axis stops, the spine becomes locked in whatever stage of erection, because the outer groove is curved.

The spine cannot be depressed in a preserved or living specimen rotated about the ridge of the coracoid by simply employing a force directly toward depression. The spine will break off sooner than go down. It can easily be depressed by twisting the spine, key-fashion; the left one clockwise, the right one contrary, at the same time that force is being applied to depress it.

The above description was made with the left side of *Pseudodoras niger* in mind. There are small modifications from this type in other species.

The pectoral spines and everything connected with them, the serration, the high development of the humeral process, frequently with spines opposed to those of the pectoral spine, the great development of the clavicle and coracoid to provide attachment for the pectoral muscles, the expansion of the clavicle to provide the elaborate groove in which the flange of the spine moves, all point to functions of the spine beyond

and in addition to that as a defensive organ. It appears as though the entire evo-
lution of the Doradidæ centered in the pectoral spine, which is well shown in Plate
XXV.

Lock of the dorsal spine.—The dorsal spine may also be locked in various positions.
It can always be erected, but depressed only when the "latch" is raised. Its articular
surface and lock involve two interneurals, that below the spine itself, and that below
the "latch," the modified spine in front of it, and the modified first ray or spine which
forms the latch. In *D. micropæus* the anterior interneural involved (Fig. 7, *E*, *INT.
I.*) consists of a median plate (*m.pl*) and lateral flanges (*f*) continuous with similar
flanges (the parapophysis) of the fourth vertebra. The tip of the first interneural in
front of its flange is expanded to form the "*Y*" element of the dorsal plate.

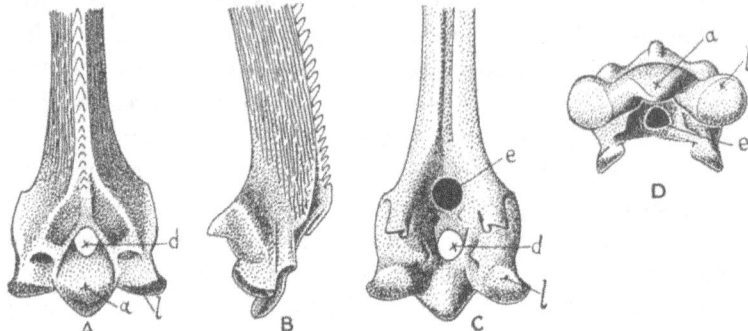

Fig. 7. The base of the dorsal spine of *Pseudodoras*; *A*, from in front; *B*, from the right side; *C*, from behind; *D*, from
below. Kennedy, del. *a*, the anterior articular surface in contact with the second interneural which passes through the opening
d; *e*, the opening through which blood vessels, nerves, etc., enter the spine; *l*, the lateral articular surfaces bearing on the lateral
flanges of the second interneural.

Behind the flange the tip of the first interneural is transformed into a two-
faceted, smooth, articular surface (*b*, Text-figs. 7 and 8) facing forward and upward.
Looked at from above and in front it resembles very much a partially unfolded young
leaflet. The two-pronged fulcrum (marked *O* and *E* or *D.S.I*), the modified first
spine of the dorsal, straddles this. The second interneural has a pair of similar
median lateral ridges which are expanded at the tip and provided with articular
surfaces (*K*, Text-fig. 8) on which the lateral ends (*l*) of the dorsal spine move. The
anterior part of middle plate of the second interneural is developed into a concave
articular surface (*c*, Figs. 8 and 9) in which the anterior, mesial end (*a*) of the dorsal
spine moves.

A finger-like prolongation from the middle process of the interneural is crooked
backward through an opening in the dorsal spine (*B*, Text-fig. 7). It meets a similar

forward-crooked process from the posterior part of the second interneural, the two uniting to form a ring by which the dorsal spine is fastened to the second interneural. As long as the spine is erect and the crooked fulcrum is down as in Text-fig. 9, *A*, *E*, and *G*, the spine is locked, erect. The spine can be depressed when the fulcrum *O* is raised as in *D* and *F*.

Fig. 8. *A*, Base of dorsal spine, and *B* to *E*, elements of the dorsal lock of *Pseudodoras niger*. The right in figures *A*, *B*, and *E*, are anterior. *A*, base of spine from the right; *B*, tip of first and second interneural from the side; *C*, tip of first interneural from in front. The tip of the second neural spine that passes through the dorsal spine at *d* is broken off. *D*, the "latch," the fulcrum, or first dorsal spine from in front. *E*, the same from the side.

In *Megalodoras irwini* a median partition or interneural septum extends down from the "*X*" element of the dorsal plate (*F* and *G*, Fig. 9). In *D. micropœus* there is no such partition (*D* and *E*, Fig. 9). There are many gradations in the development of the median partition between these two extremes.

An interneural septum below the "*X*" element of the dorsal plate is complete or nearly complete in *Centrochir crocodili*, *Platydoras costatus*, *Lithodoras dorsalis*, *Anadoras weddelii* and *grypus* and in the adult of *Pseudodoras niger*.

The septum is not complete in *Pterodoras granulosus*, *Astrodoras asterifrons*, *Acanthodoras spinosissimus*, *Amblydoras hancocki*, *Doras carinatus*, *D. punctatus*, *Opsodoras humeralis*, *Leptodoras linnelli* and *Trachydoras paraguayensis*.

Judging from the miscellaneous species in which it is complete or nearly complete, and those in which it is incomplete, this character cannot be given much weight in classification.

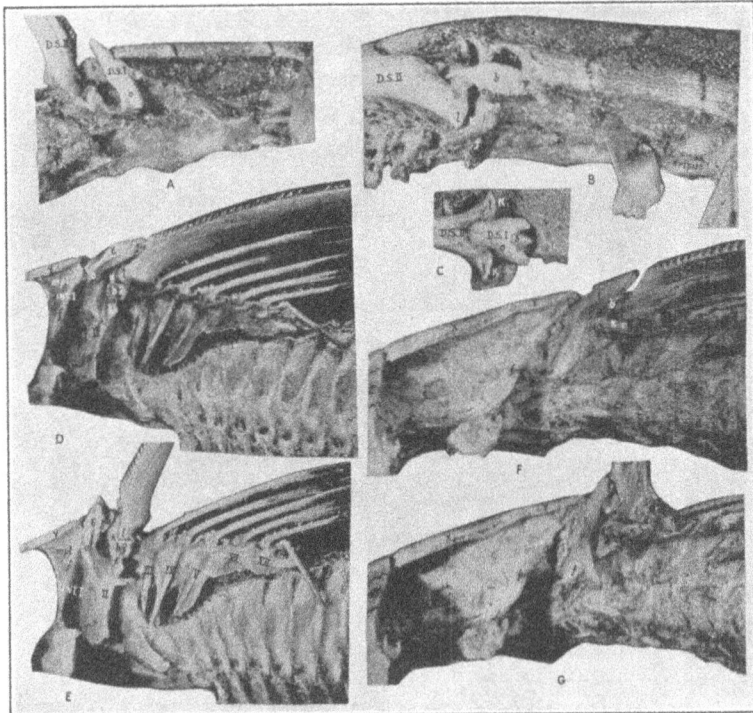

FIG. 9. *A*, Dorsal lock in *Platydoras costatus* from the side. *B*, The same three-quarter view. *C*, The same from the top. *D* and *E*, Dorsal erect and depressed in *Doras micropæus*. *F* and *G*, Dorsal depressed and erect in *Megalodoras irwini*. *b*, articular surface of the first dorsal spine; *D. S.* I, first dorsal spine; *D. S.* II, second or main dorsal spine; *E*, first dorsal spine; *f*, lateral flange of the I and II interneurals; *K*, articular surface of the lateral flange of the second interneural; *l*, lateral articular surface of the dorsal spine; *mpl*, middle plate of the interneural; *o*, first dorsal spine.

Surface plates.—Immediately above the humeral process the air-bladder is in contact with the skin to form a tympanum. The air-bladder space is separated internally from the muscled region above, either by a membranous septum or by a bony process or plate (9*a* in Plates XVIII, XX and XXIV) from the epiotic to a downward-directed process of the last element of the dorsal plate and the first rib.

The series of lateral scutes consist of one to three small bones without hooks near the middle of the tympanum, and from 16 to 40 hooked bones along the sides. The first one to three hookbearing scutes are in contact with the dorsal plate above and may or may not reach the humeral process.   The last plate is on the middle of the base of the caudal.   The plates are sometimes minute, isolated and imbedded in the skin (*Hassar*, Plate XXII; *Doras*, Plate XXVII, Fig. 4).   Sometimes the plates are very wide, nearly covering the entire sides, *Platydoras* (Plate IX), *Megalodoras*

Fig. 10.   The humeral region of a specimen of *Pseudodoras niger* from La Merced, showing the supplementary lateral scute at *X*.

(Plate VI), and *Acanthodoras* (Plate XI).   In the extreme the plates of opposite sides may meet along the middle of the back (*Acanthodoras spinosissimus*, Fig. 2, Plate XI).

Rarely, plates are developed along the middle of the back between the dorsal and the adipose (*Lithodoras dorsalis, Hemidoras hemipeltis* and *morrisi*).   In one species, *Hypodoras forficulatus*, a large regular plate covers the anterior half of the adipose (Plate IV, Fig. 1, and Plate XXV).

In an extreme case, *Lithodoras dorsalis*, dermal plates may develop at any point on the surface and, in very old, they may form a protective armor over the entire body.

The caudal fulcra are always well developed, frequently they are laminate, extending forward on the caudal peduncle.   Sometimes the fulcra become spread out forward, forming isolated plates, or plates in contact over the entire region to the adipose and anal (Plate III, Figs. 1, 5, and 6; Plate IV, various figures).

Air-bladder.—(Plate I.)   The greatest amount of variation is found in the air-bladder.   It is indeed difficult to reconcile the fact that the various types of air-bladders belong to related species. There are five general types. Of these there are various modifications and combinations only generally correlated with other structures.

The simplest, most primitive type is found in *Centrochir crocodili*, *Hoplodoras uranoscopus* and *Franciscodoras marmoratus* (*A*, *B*, *C*, Text-fig. 12). In these species the air-bladder, externally at least, is a simple subconical or subglobular bag.

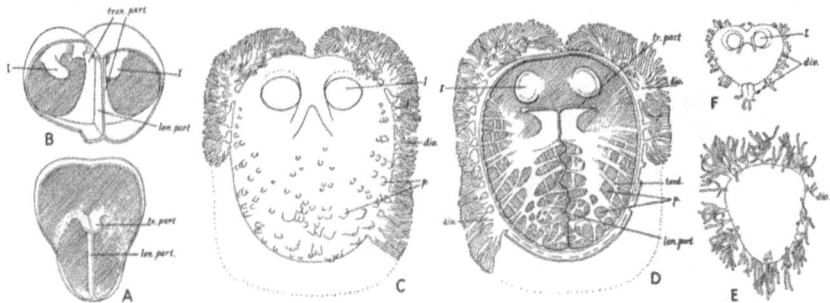

FIG. 11. Air-bladders of: *A*, *Doras carinatus*, the upper wall cut away to show the median partition in the posterior half and the partial anterior partition at its anterior end. *B*, *Doras carinatus*, transverse section, on the left immediately behind the transverse partition on the right some distance further back. Compare with Plate I, Figs. 22 and 31. *C*, *Pterodoras granulosus*, dorsal view. Compare with Plate I, Fig. 3. *D*, The same with the ventral wall cut away to show the partitions and the tendinous stays on the inner face of the dorsal wall. Compare with Plate I, Fig. 1. *E*, *Doras humeralis*, dorsal view, with many slender diverticulæ. *F*, *Megalodoras irwini*, ventral view of the young, 90 mm. long. For the adult see Plate I, Fig. 5. *div.*, diverticula; *i.*, pocket for the process of the vertebra; *lon. part.*, longitudinal partition; *tend.*, tendinous stays; *tr. part.*, transverse partition; *p.*, ventral pockets.

Internally the cavity of the bladder is partially divided into three sections. A median partition divides the posterior two thirds or three fourths into a right and left half. In front of this a partition runs partly across the bladder. A horizontal section of these partitions gives the letter "T." In front of this "T" the bladder is a cavity extending from side to side. Into this anterior cavity project the disks and prolongations of the parapophyses of the fourth vertebra connected with the Weberian apparatus. The anterior cavity of the bladder is continuous with the posterior sections along the outer wall of the bladder (Text-fig. 12 and Plate I).

Another simple type occurs in *Platydoras costatus* (Figs. 14–18, Plate I). In these species there are two simple bladders connected with each other by a very short, thin canal. The anterior bladder resembles that of *Centrochir*, the posterior (cœcum) is similar but much smaller, elongate, egg-shaped.

The third type is a modification of the second. The posterior bladder is double or split lengthwise, the two halves sometimes recurved (Text-fig. 12, *E* et al.).

A fourth type was observed once. This consists of the double air-bladder, with a third division behind the second in *Scorpiodoras heckeli* (Text-fig. 12, *D*).

The fifth type of bladder is provided with a number of radiating tubes or cœca. The walls of this type are firm, not flexible, provided with internal ridges varying greatly in number and size. In *Pterodoras* they are not unlike the ribs of a boat (Text-fig. 11, Plate I, Figs. 1–3).

The tubes vary greatly in number, number of rows, in size and in character. Some are simple, narrow, sausage-shaped diverticula around the sides of the bladder. Others are branched simply or in an arborescent fashion, or even feather or plume like, and imbedded in masses of fat (Figs. 1 to 4 and 6, Plate I).

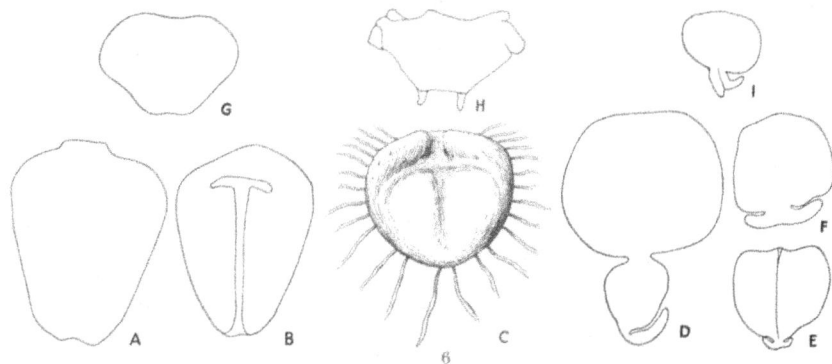

Fig. 12. Air-bladders of: *A, Centrochir crocodili* (Humboldt); *B, Franciscodoras marmoratus* (Reinhardt); *C, Hoplodoros uranoscopus* (Eigenmann and Eigenmann); *D, Sorpiodoras heckeli* (Kner); *E, Doras punctatus* (Kner); *F, Astrodoras asterifrons* (Heckel); *G, Acanthodoras spinosissimus* (Eigenmann and Eigenmann); *H, Leptodoras linnelli* (Eigenmann); *I, Hypodoras forficulatus* (Eigenmann).

The beginning of the development of cœca in the double air-bladder occurs in *Pseudodoras niger* (16, Plate I). In this species the air-bladder has circular, thin places in the side wall of the depressed bladder. It would seem that the thin walls of these disks are capable of acting bellows-like with fluctuating pressure. The walls themselves seem inflexible. The height of modification of this type is found in *Megalodoras irwini* (4, Plate I) in which the number, size and elaborateness of the branching of the diverticula reaches the truly marvelous.

The diverticula in the single air-bladder go through a similar, though not identical, gamut of development. There may be many very thin tubules in various of the species with fimbriated maxillary barbels or there may be a few larger, short out-pocketings as in *Leptodoras linnelli*. The height of elaborateness is found in *Pterodoras granulosus* (1 to 3, Plate I) and is described under that species and in species of *Opsodoras* (especially in *O. humeralis* (6, Plate I)).

20

The modifications in the second air-bladder are preserved step-wise in various species. In its simplest form it is a minute protuberance, *Doras carinatus* (Fig. 8, Plate I), next an elongate, ovate bladder with pointed end, with or without diverticula. The beginning of the double posterior air-bladder is preserved in *Lithodoras dorsalis*. In this species an external constriction partially divides the bladder. The end is divided into two lateral horns. An internal septum completely divides the rest of the posterior bladder into two unequal lateral parts.

The modifications of this type can best be described by a series of figures of bladders actually observed.

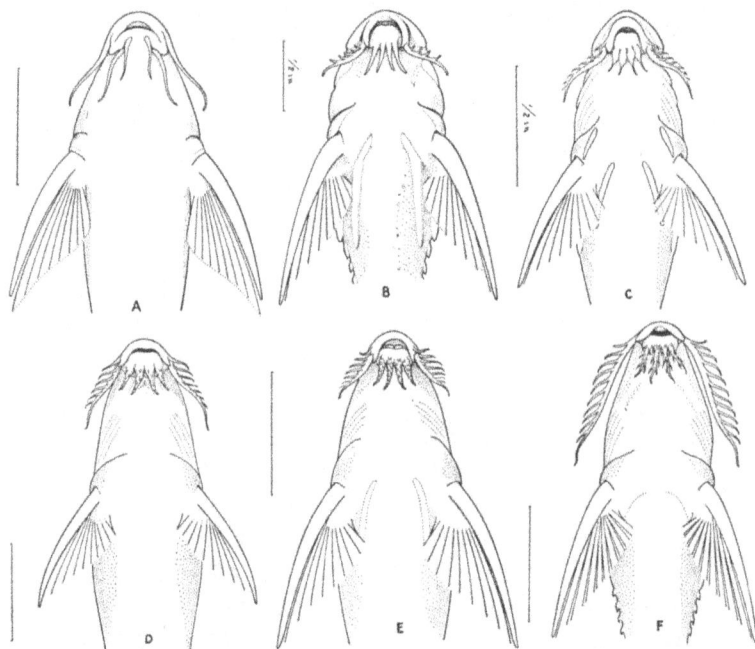

FIG. 13. Lower surfaces of the heads of different species under different magnification. The lines at the margin equal one inch unless otherwise indicated. *A, Rhinodoras d'orbignyi; B, Trachydoras paraguayensis; C. Trachydoras atripes; D, Hassar wilderi; E, Opsodoras humeralis; F, Hemidoras morrisi.*

Barbels.—The barbels are but moderately developed; there is always one maxillary and two pairs of mental barbels. In no species do the maxillaries extend any great distance beyond the tip of the humeral process. Usually they are shorter. In many species the mental barbels are warty, or carry supplementary barbels. In these the bases of the mental barbels, which are in a series close behind the lip (not paired), are frequently united by a broad basal membrane.   The height of the complication in the barbels is found in *Leptodoras linnelli* (Fig. 2, Plate XXIV).   In this species the maxillary barbel is divided beyond the bone into an inner and an outer branch.

In the species with fringed mental barbels in which the bases of the barbels are united by a membrane, the mental barbels can be bent forward to form a screen, or sieve over the mouth.

Color scheme.—The underlying scheme of coloration consists of a light band along the spines and darker bands above and below the light band.   The darker bands are continued to the tips of the caudal lobes.   The parallel bands on the caudal are frequently present when there is no indication of a longitudinal stripe elsewhere (Figs. 1 and 2, Plate IX; 1, 2 and 5, Plate XI; 1 and 4, Plate XII; 1 and 2, Plate XV). The lateral stripes are best developed in *Platydoras costatus*.   In a few species (*Hassar* and *Platydoras*) there is a black spot on the dorsal.   The most strongly marked species is *Acanthodoras spinosissimus*.   It is black, with light spots and streaks and has a banded caudal.

Weberian apparatus.—The Weberian apparatus consists of a long, narrow, oval bone, *scaphium* or *stapes*, fitting over the oval opening, *fenestra* "*a*," into the lymph spaces of the internal ear.   Between it and the *tripus*, imbedded in tendonous tissue, lies the curved *incus* or *intercalarium;* the *tripus* or *malleus* is a flat blade extending backward from the *incus* to below the lateral process of the fourth vertebra.

In some species the *tripus* remains a flat blade to its posterior end, in others it is folded upon itself from its inner edge; at a greater or lesser distance from its posterior end it comes in contact with the fourth vertebra by one or two short processes forming the pivot on which it moves.

A slender, unnamed, rod-shaped bone extends from the anterior, raised, lateral edge of the fourth vertebra backward and attaches by a fan-shaped tendon to the *tripus*.   The anterior process of the fourth vertebra arises as a thin, sinuous bone from the basal half of the neural spine.   It extends outward and more or less downward, being notched to arch over the *tripus*.

At the outer edge it is continued, or attached, to a disk or cone, fitting into an indentation in the anterior or upper wall of the air-bladder.   In contact with the bone the wall is thin, delicate.   If conical the apex of the cone is indented into the air-bladder (Plate XIX, Fig. 4, *i;* XX, Fig. 1, *i*).

The bones are so delicate, so small and so variable that I am not able to state positively that these structures are built on this plan in all the species.   The small rod from the *tripus* to the anterior edge of the fourth vertebra seems to be especially variable. The part of the process of the fourth vertebra which is hooked over the outer edge of the *tripus* is also variable.

The body of the fourth vertebra is very long, its lower surface smooth and intimately associated with the air-bladder. Sometimes (in *A. spinosissimus*) a deep median septum is developed in the mid-ventral line. The anterior end of the vertebra is always well marked, sometimes by a thickening or general enlargement,

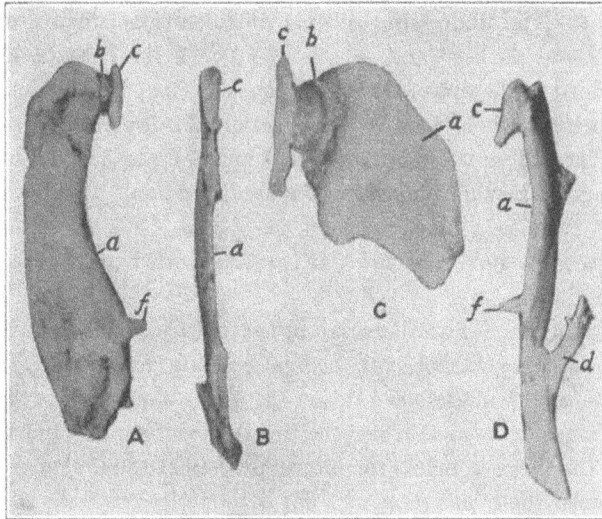

FIG. 14. The auditory, or Weberian ossicles of *Pseudodoras niger*. *A*, The right ossicle seen from below; *B*, the same from the vertebral viewpoint; *C*, the anterior end of *A* seen from above, more enlarged. The fulcrum, *f*, situated near the middle in *A* and *B*, is broken and restored in outline in *A*. *D*, the ossicles in another individual showing the supplementary bone, *d*, attached at the right; *a*, tripus or malleus, supposed to be the modified rib of the third vertebra; *b*, intercalarium or incus, supposed to be the modified neural arch of the second vertebra; *c*, scaphium or stapes, supposed to be the modified neural arch of the first vertebra. The claustrum, not shown in the photograph, is supposed to be the spine of the first vertebra. More probably the "scaphium and claustrum" represent the first neural arch and rib.

sometimes by a sharp ridge. The latter is much exaggerated in *Leptodoras*, in which it forms a deep pocket divided by a median septum. Into these pockets the anterior ends of the air-bladder fit.

DISTRIBUTION AND RADIATION OF THE DORADIDÆ.

The species of the Doradidæ belong prevailingly to the lowlands of the Amazons, ascending in the Chanchamayo to at least 2,500 feet, and in the Urubamba to 3,000 feet.

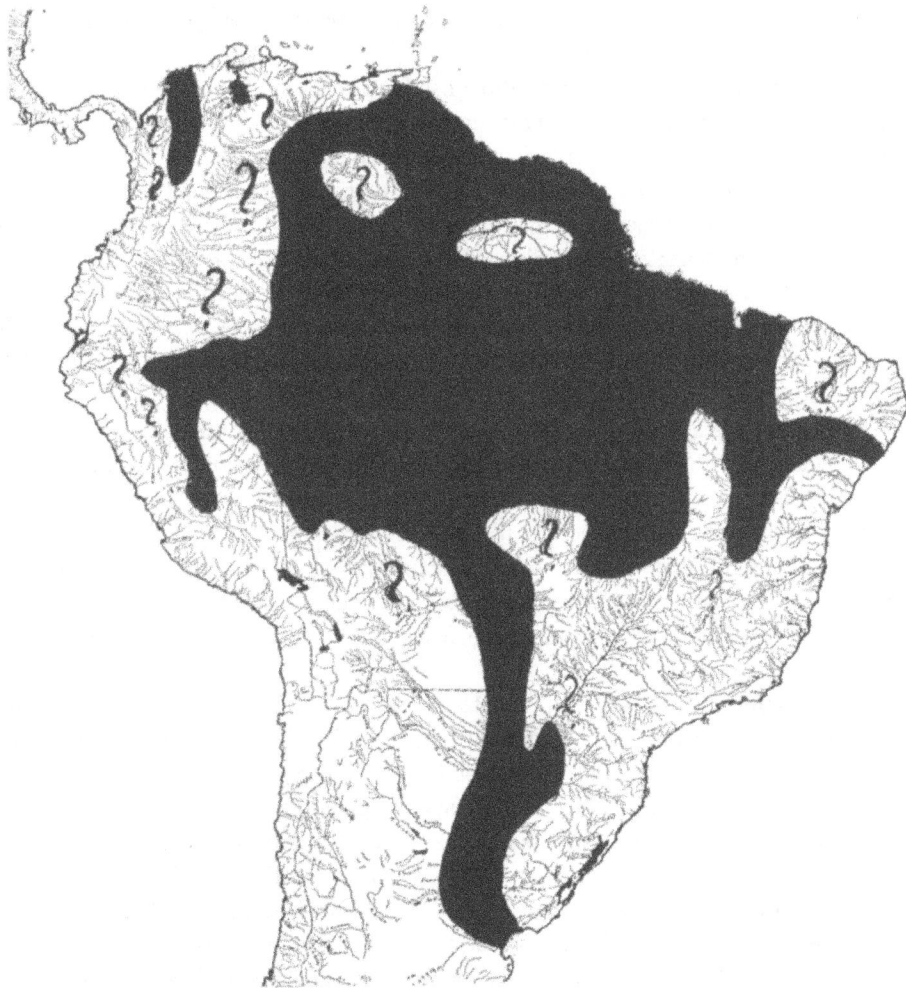

Fig. 15. Distribution of the Doradidæ. Areas in doubt marked ?.

They extend in a long pseudopodium down the Paraguay to the La Plata. The number of species that extend into the La Plata-Paraguay basin is but five, of which two, the genera *Oxydoras* and *Rhinodoras*, are peculiar to that system. Two other species extending into the La Plata system, *Pterodoras granulosus* and *Platydoras costatus*, are the most widely distributed species of the family. The fifth, *Trachydoras paraguayensis*, occurs on both sides of the divide separating the Paraguay and the Amazon. The La Plata is beyond the normal range in climate of the species and their southward extension is evidently due to the facilities for easy movement furnished by the Paraguay-La Plata river.

At that, no member of the Doradidæ has been recorded from the upper Parana. Nor do they occur in any of the rivers emptying between the La Plata and the Rio San Francisco. Their point of origin is evidently the Amazonian basin; all but *Oxydoras, Rhinodoras* and *Franciscodoras* occur in the Amazon basin. The Rio San Francisco has provided means of migration to points not otherwise accessible and is responsible for another pseudopodium-like extension of habitat. They are found throughout the San Francisco but not east of it as far as is known.

One of the species of the San Francisco, *Franciscodoras marmoratus*, with a simple air-bladder is peculiar to it. *Platydoras costatus* and *Pseudodoras niger* are the other species of the San Francisco. They are widely distributed species.

I found no members of the Doradidæ above the Kaieteur falls and on the basis of that fact I have left the highlands of the Guianas blank in the distribution map.

But one species is found west of the Andes, *Centrochir crocodili.* It is a species without strongly marked characters; it, like *Franciscodoras,* has a simple air-bladder.

The Brazilian Amazons are most abundantly supplied with members of the Doradidæ. Next comes the Marañon. As far as the evidence goes at present, the species of the genus *Doras* prevail in the Brazilian Amazons while those of its ally, *Opsodoras,* prevail in the Marañon.

As far as the evidence goes, it points to the conclusion that the Doradidæ antedate the present configuration of the country and that they were in what are now the Magdalena, San Francisco, and La Plata basins from early times.

They evidently formed part of the fauna segregated by the formation of the Cordillera of Bogota and have not had access to the Magdalena since these mountains were formed. *Franciscodoras,* the peculiar genus of the Rio San Francisco, is evidently the original member inhabiting the Rio San Francisco. *Platydoras* (and *Pseudodoras?*), found everywhere, may be part of the original fauna; but as they are not found in the Magdalena, they may be later products which have migrated into the San Francisco by way of the Tocantins, Somo and Rio Preto.

Similarly, the genera peculiar to the La Plata basin, *Rhinodoras* and *Oxydoras,* may be looked upon as the more slightly modified descendants of the original fauna. *Platydoras costatus* may also be an original inhabitant but has less claim since it would involve the supposition that it has retained its character, both in the San Francisco and the Plata, since early time. It seems clear that *Trachydoras paraguayensis,* belonging to an otherwise Amazonian genus and to a species also found in the Amazon basin, is an interloper in the Paraguay.

Centrochir of Trans-Andes, *Franciscodoras* peculiar to the San Francisco, *Rhinodoras* and *Oxydoras* peculiar to the La Plata, are of very long-standing and a composite of these and of their nearest, and at the same time most widely distributed, relatives probably presents a generalized portrait of the earliest Doradid. See *Centrochir crocodili* (Plate VIII); *Franciscodoras marmoratus* (Plate XXIX); *Rhinodoras d'orbignyi* (Plate XXVI, Fig. 5); *Pseudodoras niger* (Plate XVII); *Platydoras costatus* (Plate IX). To these should probably be added *Pterodoras granulosus* (Plate VII, *a*), generalized Doradid externally but with an incomprehensible air-bladder.

From these generalized species there have been specializations along some definite lines, the modifications of one character frequently not correlated with modifications in other characters.

Megalodoras has gone but a short distance. It has emphasized the spines of the fins. *Lithodoras* has reached the maximum in the development of the number of dermal plates, which, however, have also increased in *Hypodoras* and *Hemipeltis*. The size of the lateral scutes has gone both up and down from the normal. They have reached their minimum, not in number but in general development, in *Doras micropœus* and *Opsodora hemipeltis*. I was inclined to place a higher value on the reduction in the size of the plates than it merits. The species with reduced plates are more nearly related to different species with well-developed plates than they are to each other. The reduction of the plates has taken place independently in different species. The pairs *Doras carinatus* to *Doras micropœus* and *Opsodoras parallelus* to *hemipeltis* are evidently more closely related to each other than the last of each of these pairs with minute plates is to species of *Hassar* which have minute plates.

The upward variation in the size of the plates reached its maximum in *Acanthodoras*. In this and the nearest related forms the scutes are variously provided with spines. The caudal fulcra, well developed in the typical species, have undergone development in several independent lines. In several independent lines the anterior ones have become enlarged and cover the caudal peduncle more or less completely.

The modifications of the pectoral musculature, the granulation of the coracoid and the development of a "diaphragm" have been mentioned above.

The mouth, the dentition, the barbels and the food habits have been associated, correlated modifications, directed distinctly away from the generalized *Centrochir* and *Franciscodoras*. In a measure each genus and species is the end product of series of modifications from the generalized Doradid. Among those that have gone furthest afield are *Trachydoras* with its associated *Hemidoras* and *Doras*, and *Leptodoras* with its associated *Nemadoras*, *Hassar* and *Opsodoras*.

TABLE OF DISTRIBUTION.

The following table indicates the species not seen by the author, the species known from the types only (marked *), the length in millimeters of the largest recorded specimen and the recorded distribution of the various species.

	Largest Specimen Known.	La Plata.	Paraguay.	Bolivia.	Bahia to Rio Grande do Sul.	Rio San Francisco.	Brazilian Amazon.	Marañon.	Guianas.	Orinoco and Venezuela.	Magdalena.
I. *Megalodoras irwini*	612						×	×	×		
libertatis (not seen)†	112						×				
II. *Centrodoras brachiatus*	410						×	×			
III. *Hoplodoras uranoscopus**	530						×				
IV. *Pterodoras granulosus*	230	×	×	×			×	×	×		
*lentiginosus**	357						×				
V. *Centrochir crocodili*	210										×
VI. *Platydoras costatus*	217		×	×		×	×	×	×	?	
albomaculatus (not seen)*	85									×	
helicophilus (not seen)*	350								×		
dentatus (not seen)*	135								×		
VII. *Franciscodoras marmoratus*	360					×					
VIII. *Lithodoras dorsalis*	960						×	×	×		
IX. *Acanthodoras calderonensis* (not seen)	103						×	×			
cataphractus	170						×		×		
spinosissimus	150						×				
X. *Agamyxis pectinifrons* (not seen)*	76							×			
flavopictus (not seen)*	110							×			
XI. *Astrodoras asterifrons*	100				×		×				
XII. *Scorpiodoras heckelii*	173						×				
XIII. *Amblydoras monitor**	38						×				
hancocki	123						×	×	×		
XIV. *Anadoras regani* (not seen)*	90						×				
grypus	150						×	×			
nauticus (not seen)*	82							×			
weddellii	113			×			×	×			
*insculptus**	100						×				
XV. *Hypodoras forficulatus**	123							×			
XVI. *Physopyxis lyra**	35							×			
XVII. *Pseudodoras niger*	840					×	×	×	×		
huberi (not seen)*	400						×				
XVIII. *Oxydoras kneri*	360		×								
XIX. *Rhinodoras d'orbignyi*	160	×	×								
XX. *Trachydoras paraguayensis*	93		×	×			×				
nattereri	130?						×	×			
atripes	103							×			
trachyparya (not seen)*	93						×	×			
XXI. *Doras fimbriatus* (not seen)*	125						×				
brevis (not seen)*	125						×				
punctatus	125						×				

	Largest Specimen Known.	La Plata.	Paraguay.	Bolivia.	Bahia to Rio Grande do Sul.	Rio San Francisco.	Brasilian Amazon.	Marañon.	Guianas.	Orinoco and Venezuela.	Magdalena.
XXII. Doras eigenmanni	95		×				×				
microstomus	52						×			×	
carinatus	267						×			×	
micropœus*	270									×	
lipophthalmus (not seen)*	190						×				
XXIII. Hemidoras stenopeltis	160			?			×	×			
morrisi	147							×			
XXIV. Opsodoras hemipeltis*	143							×			
parallelus*	151							×			
humeralis	154						×	×			
orthacanthus*	133							×			
trimaculatus	77						×		×		
leporhinus	81						×			×	
ternetzi	129						×				
boulengeri (not seen)*	167						×				
stübeli (not seen)*	120								×		
morei (not seen)*	125						×				
steindachneri (not seen)*	235						×				
XXV. Hassar affinis (not seen)*†	255										
orestis	196						×				
wilderi*	215						×				
notospilus*	70								×		
XXVI. Leptodoras linnelli	209						×		×		
acipenserinus	223						×?	×			
juruensis (not seen)*	235						×				
hasemani (not seen)*	132						×				
XXVII. Nemadoras elongatus (not seen)*	105						×				
bachi (not seen)*	90						×				
TOTALS		2	5	5		3	47	26	15	2	1
							58				

† Paranahyba and Itapicuru.

Along one line, the dorsal spine, originally serrate, has become less and less serrate till it has lost all serration in *Astrodoras*. The preorbital, procumbent and smooth in the original Doradid has become serrate and pectinate, the marginal teeth becoming erect, a line of development reaching its maximum in *Agamyxis* and *Astrodoras*.

Along another series the eye has become reduced in size, a modification indicated by peculiar adjustments in the orbital bones.

Conclusion.—Members of the Doradidæ are found from Buenos Aires through the Paraguay, Madeira, the Amazons to the Guianas and from the mouth of the Rio

San Francisco and Para to an elevation of about three thousand feet at the base of the Andes in Bolivia and Peru. One member of the family, *Centrochir longispinis*, is found in the Magdalena river, across the first chain of the Cordilleras in Colombia. As far as is known, they do not occur in the rivers between the mouth of the Rio San Francisco and the Rio Plata or in the upper courses of the Rio Parana.

KEY TO THE GENERA OF THE DORADIDÆ.

a. Head depressed; width at the clavicle greater than the length of the head to the end of the bony opercle; eye, if small, far in front of the middle of the head; if large or medium, near middle of the head; mouth terminal, wide; premaxillary wide, fixed, with a band of teeth; epiotic bordered by the occipital, two nuchal plates, the supraclavicle and pterotic; process of epiotic not reaching the first rib and first hook-bearing lateral scute, except in *Pterodoras;* barbels all simple.

  b. Adipose fin continued forward as a low keel, longer than anal fin; snout smooth, preorbital bones not pectinate or serrate; caudal forked.

    c. Caudal peduncle naked above and below, or the posterior half covered with laminate caudal fulcra.

      d. A second air-bladder or cœcum; both depressed and fringed with numerous diverticula (not examined in *libertatis.*

        *. Less than 23 lateral scutes; fewer thorns on posterior face of the dorsal spine than on the anterior..............................1. **Megalodoras** Eigenmann, gen. nov.

        . Scutes about 38–40 on a side. ....2. **Centrodoras Eigenmann, gen. nov.

      dd. Only one air-bladder.

        e. Lateral scutes 18; air-bladder wider than long, with a marginal row of long, diverticula; thorns on back of dorsal spines fewer and smaller than on anterior margin.

                              3. **Hoplodoras** Eigenmann, gen. nov.

        ee. Lateral scutes 25 to 30, very narrow, strongest over end of anal; eye entirely in the front half of the head; about 20 thorns on the posterior face of the dorsal spine, nearly of equal size with those covering the opposite side or anterior face; no posterior air-bladder, the anterior depressed, margined with fimbriated diverticula on the posterior and lateral edge, a long plume-like diverticulum on the outer anterior angle, curved around to the front; a bony stay behind the tympanum....................4. **Pterodoras** Bleeker.

        eee. Lateral scutes 30. Fontanel not constricted; skull finely granular, arched; posterior hooks of the dorsal spine stronger than the anterior; three hook-bearing scutes in contact with the dorsal plate, the rest of the scutes graduate to the last, narrow; depth of the air-bladder more than half its width, without diverticula, no second air-bladder or cœcum. (See also *bb.*).......................................5. **Centrochir** Agassiz.

    cc. Caudal peduncle entirely covered with plates above and below (modified fulcra); scutes very deep, leaving only a narrow naked area along the back; three scutes in contact with the dorsal plate; posterior thorns of the dorsal spine disappearing with age; eye in middle of the head; air-bladders without diverticula, large; depth of the anterior air-bladder half its width; posterior carrot-shaped...6. **Platydoras** Bleeker.

  bb. Adipose fin high, longer than anal, not continued forward as a keel, or continued for a short distance only; dorsal spine serrate in front only; eye nearer snout than to end of opercle; caudal peduncle largely covered with fulcra above and below, increasing in size forward. Scutes about 30, about one third or a little less of the depth; caudal forked or deeply emarginate; fontanel without groove, its depression ending abruptly; anterior air-bladder large, divided by a median partition to very near the anterior end, without diverticula; no posterior bladder or cœcum.

7. **Franciscodoras** Eigenmann, gen. nov.

bbb. Adipose fin shorter than anal, not shading into a keel in front (*A. calderonensis?*).

    f. Dorsal spine serrate in front and behind, the posterior serræ stronger than the anterior; lateral scutes 16 to 20; caudal forked....................................8. **Lithodoras** Bleeker.

    ff. Dorsal spine serrate in front, not behind; caudal round or truncate.

        g. Caudal peduncle without plates above and below; nasal bone short, with less than ten spines; eye minute, about 4 in interorbital, far in advance of the middle of the head; fontanel subcircular; breast entirely covered with skin or the tips of the coracoid process granular; sides of dorsal spine and upper surface of the pectoral spine serrate, except in *A. calderonensis.* Caudal round or truncate. (Species with light stripe along the sides.)

        9. **Acanthodoras** Bleeker.

        gg. Caudal peduncle with spiny plates above and below; sides of the dorsal and pectoral spines serrate; nasal plate erect, pectinate; eye moderate, in advance of the middle of the head; dorsal and pectoral spines with spines along the side. (Species without a light line along the spines.)..10. **Agamyxis** Cope.

        ggg. Caudal peduncle with prominent laminate caudal fulcra covering about half its surface above and below; dorsal and pectoral spines with ridges on the sides.

        h. Nasal plate erect; width at clavicle greater than distance from snout to end of dorsal plate; posterior air-bladder simple or forked.......................11. **Astrodoras** Bleeker.

        hh. Nasal plate procumbent; width at clavicle about 1.25 in distance between snout and end of dorsal plate; second air-bladder heart-shaped, with a simple cœcum.

        12. **Scorpiodoras** Eigenmann, gen. nov.

fff. Dorsal spine grooved, without spines on sides, front, or back, except in *Physopyxis.*

    i. Adipose fin present; coracoid process shorter than humeral process.

        j. No plates between dorsal and adipose; caudal truncate.

        k. Granulations of coracoid and clavicle confluent into a large buckler; nasal serrate.

        13. **Amblydoras** Bleeker.

        kk. Granulations of coracoid and clavicle not confluent, a space covered with skin between the granulations; nasal not serrate............14. **Anadoras** Eigenmann, gen. nov.

        jj. A large plate over anterior part of adipose; caudal truncate; nasal serrate.

        15. **Hypodoras** Eigenmann, gen. nov.

    ii. No adipose; coracoid processes lyre-shaped, reaching ventrals, much larger than humeral processes which do not reach the first lateral scute; caudal rounded.

        16. **Physopyxis** Cope.

aa. Width at the clavicle less than the length of the head; snout conical or subconical; mouth relatively small; premaxillaries attached to the front of the maxillary, not joined to each other, not joined to the ethmoid, which is pointed and without lateral wings in front. Teeth feeble or none. Caudal forked.

l. Nuchal shields without a foramen.

    m. Barbels simple. Mental barbels not united by a membrane joining their bases. Adipose fin continued forward as a low keel, longer than anal fin; epiotic bordered as under *a* of the Key, not joined by process with the first hook-bearing scute; first nostril remote from the lip.

   n. Eye behind middle of head; caudal peduncle naked above and below. Serræ on posterior
        margin of the dorsal spine antrorse, or at right angles to the spine, less numerous and much
        more feeble than those of the anterior margin; fontanel with a bridge.
      o. Lateral scutes 16–23; rows of tentacles on roof of mouth........17. **Pseudodoras** Bleeker.
     oo. Scutes 34–40; no tentacles on roof of mouth.....................18. **Oxydoras** Kner.
  nn. Eye in middle of head; caudal peduncle entirely covered above with laminated caudal fulcra,
        nearly entirely covered below; serræ on posterior face of dorsal spine retrorse, much longer
        than those on anterior; fontanel not divided; lateral scutes about 30.

                                                      19. **Rhinodoras** Bleeker.

 mm. Maxillary barbels fringed, the bases of the mental barbels united by a membrane; eye behind
        the middle of the head; dorsal serrate on both margins; caudal forked; adipose short; epiotic
        forming part of the edge of the skull between the supraclavicle and the nuchal plate, with a
        process to the first hook-bearing scute; second air-bladder minute or absent.
     p. Opercle, preopercle and coracoid-process granular; first lateral scute large; fontanel not con-
          tinued as a groove; adipose not continued as a ridge.

                                                 20. **Trachydoras** Eigenmann, gen. nov.

    pp. Opercle, preopercle and coracoid process covered with skin. Back and ventral surface without
          plates. Adipose not continued as a ridge........................21. **Doras** Lacépède.
 ll. Nuchal shields with a foramen on each side.
    q. Maxillary barbel fringed with minute barbels, otherwise as under *ll.*
       r. A series of plates between the dorsal and adipose and sometimes between the ventrals and
            anal..22. **Hemidoras** Bleeker.
      rr. Back without scutes.
         s. Origin of ventrals nearer caudal than to snout; adipose not continued forward as a ridge.
            t. Eye normal; scutes usually well developed........23. **Opsodoras** Eigenmann, gen. nov.
           tt. Eye elongate; snout very long and slender; humeral process large and rounded.

                                              24. **Hassar** Eigenmann and Eigenmann.

        ss. Slender, origin of ventrals nearer snout than to caudal; air-bladder very short, fitted into
              cupped downward processes of the coalesced vertebræ; adipose continued forward as
              a keel..25. **Leptodoras** Boulenger.
   qq. Maxillary barbels simple; mental barbels connected by a basal membrane; adipose short.

                                             26. **Nemadoras** Eigenmann, gen. nov.

Genus I. MEGALODORAS Eigenmann, gen. nov.

Type, *Megalodoras irwini* Eigenmann.

Lateral plates few; adipose fin continued forward as a keel; head depressed, snout not conical; serræ of posterior margin of dorsal spine fewer and weaker than the anterior, obsolete with age; eye in middle or near middle of the head; suborbital chain (Fig. 4, Plate 6) consisting of three bones, the anterior in contact with the lower surface of the preorbital; the second borders about half the eye, is much wider than the posterior and has small spines at the upper border, its anterior end in contact with the first and with the preorbital; preorbital large, granular along its upper margin, in contact with the premaxillary and maxillary in front; mouth subterminal; occipital region somewhat roof-shaped, with or without median groove.

Air-bladder (of *Megalodoras irwini*, Fig. 4, Plate I and Text-fig. 12) much depressed, in two parts. The anterior part heart-shaped, nearly as wide as long, with about a dozen tubular cœca or diverticula in a row from the outer posterior angle around the sides to near the center in front; the cœca largest behind, smallest in front, all of them profusely branched; a few much smaller cœca in a row behind the outer row in front; the posterior bladder as long as the anterior, a constriction between the two; much narrower than the anterior and tapering backward. About seven lateral diverticula along the sides of the second bladder, decreasing in size backward and branched like the posterior ones of the anterior bladder. The chief features of the bladder are developed in a specimen 90 mm. long; but the posterior bladder shorter.

The air-bladder has not been examined in *libertatis*.

KEY TO THE SPECIES OF MEGALODORAS.

a. Lateral scutes, 17 or 18.

    b. Eye 2.5 to 3 in the interorbital.

        c. Fontanel narrow, not continued as a groove; dorsal spine smooth behind; scutes 2 + 17; fins spotted; lower half of sides mottled and spotted. 700 mm. over all.....1. **irwini** Eigenmann.

        cc. Fontanel continued to the middle of the occipital as a wide shallow groove, to the end of the dorsal plate as a narrow groove; eye nearly 3 in the interorbital; scutes 15 to 17. Gill-opening extending to middle of the pectoral spine; dorsal spine with few serræ behind.

                                2. **irwini** Eigenmann, spec. nov.

    bb. Eye 1.6 in the interorbital; scutes 18; dorsal spine strongly serrate on both margins. Fontanel continued to the end of the dorsal plate.................................3. **libertatis** Ribeiro.

1. Megalodoras irwini Eigenmann.

Plate 25, Fig. 1.

T. R. S. No. 2567; type, skin; Kartabo, Bartica District, British Guiana, September 11, 1919; William Beebe.

Measurements in life: length, 612 mm.; length entire, 700 mm.; width, 170 mm.; depth, 110 mm.; head, 140 mm.; interorbital, 48 mm.; weight, 9 pounds; D. I, 6; A. 12½; scutes 2 + 17; width of head equal to its length; eye 3.5 in snout, 6.5 in head, 2.6 in interorbital; dorsal spine 110 mm. long, striate on the sides, smooth behind, with about 40 serræ on the anterior margin, increasing slightly toward the tip; pectoral spine 155 mm. long, striate on the sides with about 50 serræ on the anterior margin, 38 on posterior; humeral process narrow, two thirds as long as pectoral spine; fontanel a narrow slit, ending abruptly over middle of eye, not continued as a groove, not constricted; two angular plates in the tympanum, in a line with the spines of the plates; the third to ninth plates highest, a little more than snout and eye, the hooks increasing in size to above the anal. The following is taken from the field notes: Ground color pinkish buff, becoming clear cream buff on the under side of the head. Top of the head indistinctly mottled with black. Side plates, ventral, anal and caudal fins irregularly mottled or spotted with black. Pectoral fins almost entirely black, pinkish buff at base. Interplate space of the back black, as is also the skin just back of the side of the top of the head, and anterior to the large side plates. Belly indistinctly mottled with black, the mottling more concentrated at the sides.

The stomach of this fish was entirely empty, while the intestine had seven specimens of the common large brown river snail, evenly spaced along its length.

The Creole name of this fish is "Key-way-mamma," which translated means "mother of the snails," in allusion to its habit of feeding upon snails. The fish are not eaten by the natives, their flesh being too tough and not considered wholesome. This specimen was captured upon a hook.

This large specimen was originally given a distinct name. No. 16168 intermediate in size, 395 mm., shows the Guiana specimen to be but an over-large specimen of *M. irwini* and the name proposed for it has therefore been suppressed.

2. **Megalodoras irwini** Eigenmann, spec. nov.

Plate 1, Fig. 4; Plate 3, Fig. 3; Plate 6, Figs. 1–4; Plate 25, Fig. 2;
Text-fig. 11, F.

I. U. M. 15427, 5, 90, 220, 278, 300 and 315 mm. The 278 mm. specimen, the type. Iquitos, Allen 1920 and Morris 1922.

I. U. M. 16168, 2, 275 and 395 mm. Santarem market, 1924. Carl Ternetz.

Head 4; depth 4–4.5; D. I, 6; A. 12 or 13; scutes 2 + 15 to 17; width of head equal to its length; interorbital 2.75–3; eye equidistant from snout and end of opercle; its diameter 8 in the head, 3 in the interorbital; width of mouth about equal to the interorbital; teeth minute but firm, in well-developed narrow bands; snout depressed; profile along median line to occipital straight; fontanel continued as a groove to the dorsal; opercle, suborbital, and nasal granular; head granular to near the anterior nostrils; maxillary barbels in the three larger extending beyond tip of humeral process; outer mental barbels reaching beyond base of pectoral spine or entire base of pectoral fin, the inner mental barbel about half as far; dorsal spine about two-thirds as long as the pectoral spine, about equal to the length of the head less half the opercle, its sides striate, serrate in front (32 to 35) and behind (22 to 25); pectoral spine very strong, considerably longer than the head, its upper and lower surfaces striate, its anterior and posterior edges very strongly serrate; humeral process very rough, reaching to or beyond the middle of the pectoral spine; two small free plates above the humeral process, in a line with the first four hooks; the first lateral plate in contact with the dorsal plate and humeral process; the anterior lateral plates crowded, those over the anal wide, the spines increasing in size to those above the anal; surfaces of the plates above the ventrals spinulose, the spinules decreasing in number to the caudal peduncle, the plates on the caudal without spinules, but with a prominent median spine; a hard keel forward from the adipose; many caudal fulcra, but no plates on caudal peduncle above or below. Skin of back and sides with hard papillæ; a narrow light band from the dorsal to the tip of the upper caudal lobe, a wavy light band from above the eye, following the edge of the skull then back along the upper half of the lateral plates; a similar band on the humeral process and back on the lower half of the plates, the two lateral bands, irregular, with interruptions and unions; lower surface with spots decreasing in size to the middle of the belly; fins spotted, the pectoral dark anteriorly.

3. Megalodoras libertatis (Ribeiro).

Plate VII.

Doras libertatis Ribeiro, Comm. Linhas. Telegr. Estrat. Matto Grosso as Amazonas Annexo 5, 1912, p. 20 (Manaos).

Scutes 18; head 4, depressed forward, mouth equal to distance between the outer borders of the nasal crests; maxillary barbel beyond basal third of pectoral; anterior nares on lips; groove of fontanel to end of occipital process; low, radiating ridges from center of scutes to the margin of the scutes where they end in a spiny tubercle; dorsal spine reaching the 10th scute; adipose small, over posterior part of the anal; two rows of contiguous spots from the snout to the base of the upper caudal lobe; a more or less interrupted band from the snout, through eyes, along the spines to the caudal; another series, less distinct, from the angle of the mouth to the ventrals; spots on the belly. After Ribeiro. This species is known from the type only. 112 mm. long in the museum at Rio de Janeiro.

Genus II. CENTRODORAS Eigenmann, gen. nov.

Type, *Doras brachiatus* Cope.

Eye in middle of head; fontanel not constricted, not continued as a groove; three scutes in contact with the dorsal plate, rapidly graduate to about the 6th, then narrow; dorsal spine with antrorse hooks in front, with straight to retrorse hooks behind; bones of the head striate. Width a trifle more than head; plates very numerous, 38–40; air-bladder cardiform, depressed, with numerous, fimbriated or tufted diverticula; a narrower second bladder or cœcum with very profusely tufted lateral diverticula which are brittle.

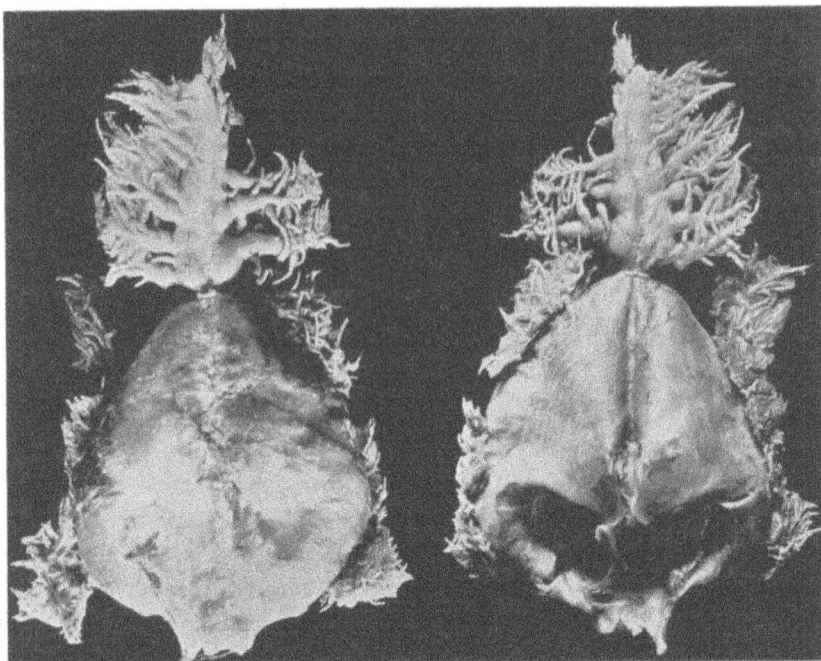

Fig. 16.　Air bladder of *Centrodoras brachiatus* (Cope).　Left from above, right from below.

4. Centrodoras brachiatus (Cope).

Doras brachiatus Cope, Proc. Acad. Sci. Phila., 1871, p. 270 (somewhere between mouth of Rio Negro and the Huallaga); Eigenmann and Eigenmann, Occasional Papers Cal. Acad. Sci., I, 1890, p. 234; Ribeiro, Peixes, IV, 1911, p. 216.

Rhinodoras amazonum Steindachner, Sb. Ak. Wiss. Wien, LXXI, 1875, p. 141, Pl. II, 1875 (Amazon near Teffé); Ribeiro, Peixes, IV, 1911, p. 197, Fig. 94.

Oxydoras amazonum Eigenmann and Eigenmann, Proc. Cal. Acad. Sci., 2d Ser., I, 1888, p. 159; Fisher, Ann. Carnegie Mus., XI, 1917, p. 421 (between Santarem and Pará).

Known from the types of *C. amazonum* and *C. brachiatus* the specimen recorded by Fisher and 16166 and 16167, 367 and 410 mm.　Santarem market.　Carl Ternetz.

Genus III.　Hoplodoras Eigenmann, gen. nov.

Type, *Doras uranoscopus* Eigenmann and Eigenmann.

This genus is very similar to *Megalodoras* in all outer appearances, but the air-bladder differs.　The air-bladder is single, wider than long and with a single series of simple cœca or diverticula.　As in *Megalodoras* the adipose fin is prolonged forward as a keel and the lateral scutes are large, few in number.

5. Haplodoras uranoscopus (Eigenmann and Eigenmann).

Plate 12, Figs. 1 and 2, Text-fig. 12C.

Doras uranoscopus E. & E., Proc. Cal. Acad. Sci., 2d Ser., I, 1888, p. 159; Occasional Papers Cal. Acad. Sci., I, 1890, p. 228.

Known from the type in the Mus. Comp. Zoöl.; a specimen 530 mm. long from Lake Hyanuary.

Caudal fulcra laminate, the first a plate.

Snout 57 mm.; eye 19 mm.; postorbital part of head 50 mm.; distance from snout to dorsal fulcrum 185 mm.; length of dorsal 49 mm.; distance from end of dorsal to end of adipose 136; distance from adipose to end of caudal lobe 160 mm.

No groove on dorsal plate; a groove from the fontanel to the dorsal plate, interrupted at base of occipital process.

Head greatly depressed, the eyes directed upward more than sideways; fins with many small spots; scutes 18; dorsal spine serrate in front, with a few hooks on its posterior margin; gill-opening extending to a point midway between upper angle of gill-opening and eye.

Genus IV. PTERODORAS Bleeker.

Pterodoras Bleeker, Nederl. Tijdsch. Dierk., I, 1863, pp. 16 and 86.

Type, *Doras granulosus* Valenciennes.

Eye small, far in advance of middle of the head; palatine bone very small; the first suborbital rudimentary, the second small, attached for its entire length along the lower surface of the thick preorbital; preorbital in contact with the premaxillary and maxillary; adipose fin prolonged forward as a keel; dorsal spine strongly serrate on both margins; lateral scutes low, 22 + 29, directly continuous (in line) with those on the pseudotympanum; caudal forked; caudal peduncle naked above and below; head depressed, snout broad; fontanel continued as a groove to the tip of the occipital process.

No posterior air-bladder (5 and 6, Plate 7); anterior air-bladder heart-shaped, with a wide fringe of tubules along the sides; a plume-like diverticulum with a double series of tubules from the anterior outer angle and curving around the anterior end, the tips of the two feather-like structures meeting near the middle in front; 3 or 4 series of short wart-like prolongations on the upper surface between the lateral diverticula, gradually reduced to a single series on the side; a bony stay or septum from the temporal plate to the process of the dorsal plate (9a, Figs. 3, 4, 5, Plate 7), separating the musculature above from the cavity in which the air-bladder lies.

21

Very similar to *Megalodoras*, differing in having narrower, more numerous lateral scutes, a very different air-bladder, and a bony stay from the epiotic, behind the tympanum (toward vertebræ). This is the only genus with simple barbels having a bony stay connecting the epiotic with the first spine-bearing scute.

6. Pterodoras granulosus (Valenciennes).

Plate 1, Figs. 1–3; Plate 3, Fig. 6; Plate 8, Figs. 1–5. Text-fig. 17.

Doras granulosus Valenciennes, in Humboldt, Rec. d'Observ. de Zoöl. et d'Anat. Comp., II, 1811, p. 184; Eigenmann and Eigenmann, Occasional Papers Cal. Acad. Sci., I, 1890, p. 229 (Arary?, Uruguay; Buenos Ayres; Serpa); Eigenmann, Repts. Princeton Univ. Exp. Patagonia, III, 1910, p. 392; Mem. Carnegie Mus., V, 1912, p. 185; Fisher, Ann. Carnegie Mus., XI, 1917, p. 419 (Santarem; Rio Mamoré).

Pterodoras granulosus Bleeker, Nederl. Tijdsch. Dierk., I, 1863, p. 15 (name only); Silures de Suriname, 1864, p. 36 (Surinam).

Doras maculatus Valenciennes, in d'Orbigny, Voy. Am. Mer., V, II, 1847, p. 7, Pl. 5, Fig. 3; Cuvier and Valenciennes, Hist. Nat. Poiss., XV, 1840, p. 281 (Buenos Ayres); Müller and Troschel, in Schomburgk, Reisen, III, 1848, p. 629; Steindachner, Denkschr. Akad. Wiss. Wien, XLI, 1879, p. 47 (Rio de la Plata); Eigenmann and Eigenmann, Proc. Cal. Acad. Sci. (2), I, 1888, p. 150 (Arary?); Occasional Papers Cal. Acad. Sci., I, 1890, p. 229 (Arary?; Uruguay; Buenos Ayres; Serpa); Ribeiro, Peixes, IV, 1911, p. 214.

Doras murica (ex Natterer, MS.) Kner, Sb. Akad. Wiss. Wien, XVII, 1885, p. 129 (Cujaba).

Doras muricus Günther, Catalogue, V, 1864, p. 202 (Demerara?).

Specimens Examined.

9836, I. U. M. 1, 127 mm.   Asuncion, Paraguay.   J. D. Anisits.
16175, I. U. M. 3, 213 to 231 mm.   Belem (Pará) Fish market.   Carl Ternetz.   May 1924.
15659, I. U. M. 2, 95–117 mm.   Marañon below Pastaza.   Allen.   Oct. 1920.
15846, I. U. M. 2, 203–216 mm.   R. Ucayali, Contamana.   Allen.   July 1920.
15847, I. U. M. 14, 87–182 mm.   Iquitos.   Allen.   Sept. 1920.
15848, I. U. M. 4, 150–200 mm.   R. Puinahua, mouth of R. Pacaya.   Allen.   Sept. 1920.

Head 4 to 4.33; D. I, 6; A. 13; scutes 1 or 2 + 23 to 28; width at clavicle a little over 3 in the length; eye 7 in the head, 2.1 in the snout, 3 in interorbital in the young, respectively 10, 7.5, 4.5 in the old; distance between the nasal openings from each other at least twice as great as the anterior from the lip; mouth terminal, broad, the upper jaw a little the longer, teeth moderate; gill-rakers minute; maxillary barbel reaching about to tip of humeral process.

Profusely spotted.

The suborbital ring of this species points to an ancestry with a much larger eye than the present.

It is very desirable that the air-bladder of specimens from La Plata be examined.

7. Pterodoras lentiginosus (Eigenmann).

Text-figs. 17 and 18.

Doras lentiginosus Eigenmann, Ann. Carnegie Mus., XI, p. 401, 2 figures, 1917. (Santarem; Manaos.)
7049, C. M. paratype, 250 mm. Haseman.

This species is known from two specimens in the Carnegie Museum. It differs but slightly if at all from *granulosus*, the head being slightly longer.

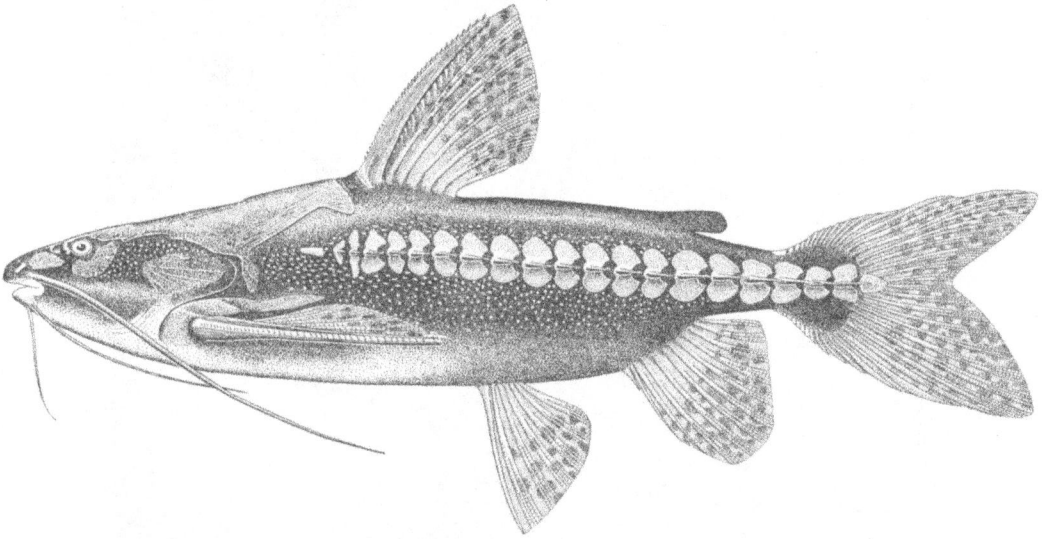

FIG. 17. *Pterodoras lentiginosus* (Eigenmann). From the type in the Carnegie Museum. Kennedy, del.

Genus V. CENTROCHIR Agassiz.

Centrochir Agassiz, Gen. et Spec. Pisc. Bras., 1829, p. 14.

Type, *Doras crocodili* Humboldt.

Width at clavicle more than length of head; lateral plates numerous, the first two or three in contact with the dorsal plate, graduate to the last, each with a simple, central hook; adipose fin continued forward as a keel; head not depressed, arched behind the fontanel, the bones finely granular; snout depressed, conical, mouth narrow, terminal; teeth well developed, in bands; serration on posterior margin of the dorsal stronger than that on anterior; fontanel continued or not continued as a groove; eyes moderate, lateral, in center or a little behind center of head; anterior nostril near the lip; caudal forked; preorbital overlying and in contact with the palatine, not reaching the maxillary and premaxillary in front, its free margin feebly but regularly notched; first suborbital under the preorbital, the second and third of about equal width, entirely in orbital border.

The air-bladder deep, heart-shaped, without diverticula, no second air-bladder (4, Plate 8; 14, Plate 2).

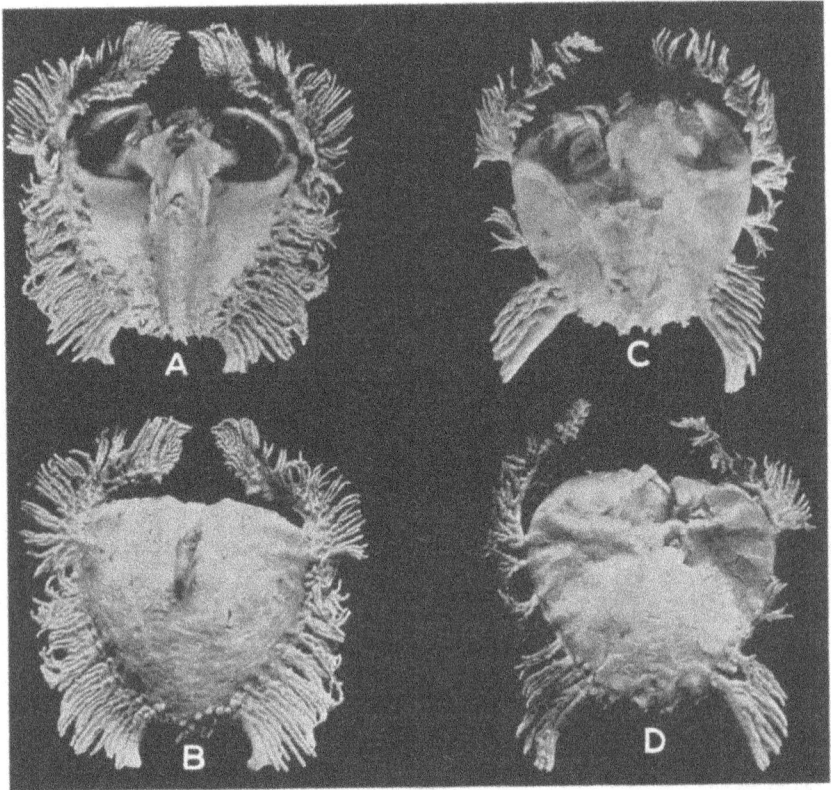

Fig. 18. Air bladders of *Pterodoras. A, lentiginosus*, dorsal view. *B, lentiginosus*, ventral view. *C, granulosus*, dorsal view. *D, granulosus*, ventral view.

Except on the obvious external characters this description is based on *C. crocodili*.

In the absence of the second air-bladder this genus differs from its nearest relatives.

Habitat: Magdalena and Amazon Basins.

KEY TO THE SPECIES OF CENTROCHIR.

a. Depth about 4.2; scutes about 30; opercle granular; caudal with parallel stripes; dorsal and ventral surfaces of the caudal peduncle partly covered by the fulcra..............8. **crocodili** (Humboldt).

8. Centrochir crocodili (Humboldt).

Plate 3, Fig. 4; Plate 5, Figs. 5 and 6; Plate 9, Figs. 4 and 5; Text-fig. 12.

Doras crocodili Humboldt, Rec. d'Observ. de Zoöl. et d'Anat. Comp., II, 1811, p. 181; Eigenmann, Mem. Carnegie Mus., IX, 1922, p. 46 (Soplaviento; El Banco; Puerto Wilches; Puerto Berrio).

Doras longispinis Steindachner, Fisch-Fauna des Magdalenen stromes, 1878, p. 23, Pl. IV, Fig. 2, Pl. V, Fig. 1 (Magdalena Basin).

Specimens Examined.

13555, I. U. M. 154–209 mm. El Banco, Rio Magdalena. Eigenmann.

13556, I. U. M. 141–194 mm. Puerto Berrio, Rio Magdalena. Eigenmann.

13557, I. U. M. 115–210 mm. Soplaviento, Rio Magdalena. Eigenmann.

Genus VI. PLATYDORAS Bleeker.

Doras Lacépède, Hist. Nat. Poiss., V, 1803, p. 116 (*carinatus* and *costatus*).

Doras Cuvier and Valenciennes, Hist. Nat. Poiss., XV, 1840, p. 267 (*carinatus* and *costatus*).

Platydoras Bleeker, Nederl. Tijdsch. Dierk., I, 1863, pp. 16 and 86 (*costatus*).

Type, *Doras costatus* Linnæus.

Width at clavicle longer than head, over 3.6 in the length; lateral scutes numerous, covering most of the sides over the anal and forward; in contact with modified caudal fulcra above and below on the caudal peduncle, the three first in contact with the dorsal plate. Adipose fin continued forward as a keel; head not much depressed, arched in the occiput; snout moderately depressed; mouth subterminal, the lower jaw a little the shorter; teeth well developed, in bands, those of the outer row enlarged in the young; serration on posterior margin of the dorsal spine weaker than on the anterior, disappearing with age; bones of the head, opercles, nasals and suborbital pitted-granular; opercle striate; eye lateral, in about the middle of the head; anterior nostrils close to the lips; fontanel without constriction; caudal emarginate; preorbital in contact with the maxillary and premaxillary in front; the first suborbital between the preorbital and the second suborbital; second suborbital little wider than the third, the two forming the lower margin of the orbit.

Air-bladder double (3, Plate 2; 2, Plate 9), the anterior heart-shaped, with a median longitudinal groove, without any cœca; posterior bladder subconical, with a rounded base, without constriction or cœca.

Habitat: San Francisco, Paraguay, Amazon, and Orinoco basins; Guianas. The most widely distributed genus.

KEY TO THE SPECIES OF PLATYDORAS.

a. A conspicuous light band from above the eyes along the middle of the sides through the caudal; scutes 28–31, leaving but a narrow naked area along the back; maxillary barbel reaching second fourth of pectoral spine or tip of humeral process....................................9. **costatus** (Linnæus).

aa. Sides plain or spotted.

   b. Dark brown, rows of large white spots above and below the lateral line; smaller white spots on belly
      and caudal; dorsal black with a few large white spots; maxillary with black and white rings; outer
      surface of humeral process with two series of spines, reaching fourth lateral scute; 29 lateral scutes;
      maxillary barbel reaching pectoral..............................10. **albomaculatus** (Peters).

   bb. Uniform blackish; dorsal fin white, its middle black; base of anal and two posterior rays white;
      humeral process without spines, extending to last third of pectoral; 32 to 34 lateral scutes, those
      on tail only half as deep as the tail; maxillary barbel reaching middle of pectoral spine; pectoral
      reaching ventral..11. **helicophilus** (Günther).

   bbb. Brown, lighter below; a dark brown spot behind each lateral hook; caudal lobes with dusky bands;
      humeral process reaching second lateral scute, with two series of spines; 31 scutes; maxillary
      barbels to last third of pectoral spine........................12. **dentatus** (Kner).

9. Platydoras costatus (Linnæus).

Plate 1, Figs. 5 and 14; Plate 3, Figs. 1 and 2; Plate 5, Figs. 1 and 2;
Plate 9, Figs. 1–3.

Silurus costatus Linnæus, Syst. Nat., ed. 12, I, 1766, p. 506.

Cataphractus costatus Bloch, Ausl. Fische, VIII, 1794, p. 82, Pl. 376.

Doras costatus Lacépède, Hist. Nat. Poiss., V, 1803, p. 116, part (South America); Cuvier and Valenciennes,
   Hist. Nat. Poiss., XV, 1840, p. 268 (Guiana); Castelnau, Anim. Am. Sud. Poiss., 1855, p. 48 (Amazon);
   Günther, Cat. Fish. Brit. Mus.,V, 1864, p. 201 (British Guiana; River Cupai); Eigenmann and Eigen-
   mann, Proc. Cal. Acad. Sci. (2), I, 1888, p. 161 (Rio Preto; Rio Puty; San Gonçallo; Xingu Cascade;
   Obidos; Gurupa; Teffé); Occasional Papers Cal. Acad. Sci., I, 1890, p. 231; Perugia, Ann. Mus. Genova
   (2), 1891, p. 34 (Villa Maria, Paraguay); Kindle, Ann. N. Y. Acad. Sci., VIII, 1895, p. 251 (Trocera
   on the Tocantins); Eigenmann and Kennedy, Proc. Acad. Nat. Sci. Phila., 1903, p. 500 (Paraguay);
   Eigenmann, Ann. Carnegie Mus., XXXI, 1907, p. 116 (Corumba; Laguna Ipacarai); Repts. Princeton
   Univ. Exp. Patagonia, III, 1910, p. 393; Ribeiro, Peixes, IV, 1911, p. 210; Eigenmann, Mem. Carnegie
   Mus., V, 1912, p. 185 (Twoca Pan and Gluck Island, British Guiana; Calobozo); Fisher, Ann. Carnegie
   Mus., XI, 1917, p. 419 (Maciél, Rio Guaporé; San Joaquin, Bolivia; Santarem; Corumbá; Puerto
   Suarez; Rio Jaurú, Paraguay Basin).

Platydoras costatus Bleeker, Nederl. Tijdsch. Dierk., I, 1863, p. 16 (name only); Silures de Suriname, 1864,
   p. 38 (Surinam).

Doras cataphractus (not of Linnæus) Schomburgk, Fishes Brit. Guiana, I, 1841, p. 158 (Rio Negro).

Doras armatulus Müller and Troschel, in Schomburgk, Reisen, III, 1848, p. 629 (Rupununi; Awaricuar).

Specimens Examined.

16170, I. U. M. 99, 118 and 132 mm.   Porto Nacional, Rio Tocantins.   Feb. 1924.   Carl Ternetz.

16171, I. U. M. 184 mm.   Peixe, Rio Tocantins.   Carl Ternetz.

16187, I. U. M. 138 to 155 mm.   Santarem market.   Sept. 1924.   Carl Ternetz.

17041, I. U. M. 2, 185 and 217 mm. to end of middle caudal rays; Lake Rogoagua, Bolivia.   Pearson.

12030, I. U. M. 1, 100 mm.   Twoca Pan, Brit. Guiana.   Wm. Grant.

12031, I. U. M. 1, 194 mm.   Gluck Island, Brit. Guiana.   Eigenmann.

15874, I. U. M. 3, 59–81 mm.   Yarinacocha, Peru.   Allen.   Sept. 1920.

15875, I. U. M. 5, 77–81 mm.   Rio Morona, Peru.   Allen.   Oct. 1920.

10139, I. U. M. 1, 88 mm.   Corumba; Paraguay; Anisits.

D. I, 6; A. 11; scutes 2 or 3 + 28 to 30.   Eye 4 to 6 in the head, 1.5–2.75 in the snout, 1.3 to 2.75 in the interorbital; bones of head granular; mouth terminal, rather narrow; teeth well developed, in narrow bands, the outer row of the upper jaw enlarged in the young; maxillary barbel to the tip of the humeral process.

A pair of lateral white bands, diverging from the fontanel, passing along the hooks of the scutes to the tip of the middle caudal rays; dorsal with a black spot near the tips of the anterior rays; a pair of black bands along the caudal lobes.

Pectoral spine and innermost rays white, the middle rays black; anal and ventrals with a large black spot; a light line forward from dorsal spine.

10. Platydoras albomaculatus (Peters).

Doras albomaculatus Peters, Mb. Ak. Wiss. Berlin, 1877, p. 470 (Calobozo).

Known only from the types, two specimens, the larger 70 mm. to base of caudal; in the Berlin Museum.

11. Platydoras helicophilus (Günther).

Doras helicophilus Günther, Proc. Zoöl. Soc., London, XXXVII, 1868, p. 229 (Surinam).

Known only from the types, three specimens, 350 mm. long, in the British Museum.

12. Platydoras dentatus (Kner).

Doras dentatus Kner, Sb. K. Akad. Wiss. Wien, XVII, 1855, p. 118, Pl. III, Fig. 3 (Surinam).

Known from the type, about 135 mm. long, in the Vienna Museum.   It is possibly the common *P. costatus*.

Genus VII.   FRANCISCODORAS Eigenmann, gen. nov.

Type, *Doras marmoratus* Reinhardt.

Adipose fin high, longer than anal, not continued as a keel, or for a very short distance only; dorsal spine serrate in front only; eye in front of middle of head; caudal peduncle nearly completely covered with modified fulcra above and below, the fulcra not in contact with the low lateral plates; fontanel not continued as a groove behind the middle of the head; caudal deeply forked or deeply emarginate; breast covered with skin; scutes large, heavy toward the end of the series.

In the general appearance, the nature of the skull, the single air-bladder, the anterior scutes, etc., this genus resembles *Centrochir*.

13. Franciscodoras marmoratus (Reinhardt).

Plate 3, Fig. 7; Plate 22, Fig. 1; Text-fig. 12, B.

Doras marmoratus (Reinhardt, MS.) Lütken, Dan. Vidensk. Selsk. Skr., 1874, p. 30 (Rio das Velhas); id., l.c., 1875, p. 146, Pl. I, Fig. 1 (Rio das Velhas); Steindachner, Sb. Ak. Wiss. Wien, LXXI, 1875, p. 147, Pl. IV; Eigenmann and Eigenmann, Occasional Papers Cal. Acad. Sci., I, 1890, p. 237; Ribeiro, · Peixes, IV, 1911, p. 204, Pl. XXXVIII, Fig. 1; Fisher, Ann. Carnegie Mus., XI, 1917, p. 420 (Penedo; Cidadedo Barra; Joazeiro).

As far as is known this species is confined to the Rio San Francisco. I have examined some of the specimens in the Carnegie Museum.

Head 4.4; D. I, 6; A. 13; scutes 31 to 32, two scutes in contact with the dorsal plate; eye 2.25 to 2.5 in snout, 6 to 7 in head, 2.33 in interorbital.

Genus VIII. LITHODORAS Bleeker.

Lithodoras Bleeker, Nederl. Tijdsch. Dierk., I, 1863, p. 84.

Type, *Doras dorsalis* Cuvier and Valenciennes.

Width at clavicle greater than length of head, about 3.5 in the length; scutes few, 18 to 20, very narrow, with a single hook, the first two in contact with the dorsal plate; adipose fin not continued as a ridge; snout truncate, not depressed, the occiput roof-shaped; mouth terminal, the teeth well developed, in bands; bones almost smooth; fontanel continued as a groove to the end of the nuchal plate; serræ on the posterior margin of the dorsal larger than those of the anterior; eye large, lateral, its center a little in advance of the middle of the head; suborbital bone large, reaching the premaxillary and maxillary in front, with divergent ridges ending in points along its free margin. Second suborbital narrower than the third, the first appearing as a granular ridge below the preorbital; preopercle smooth; all the naked parts of the body becoming more or less covered with granular scales with age.

14. Lithodoras dorsalis (Valenciennes).

Plate 2, Fig. 7; Plate 9, Fig. 6.

Doras carinatus Valenciennes, in Humb. Rec. d'Observ. Zoöl. et d'Anat. Comp., 1811, II, p. 184 (not *Silurus carinatus* L.).

Doras dorsalis Cuvier and Valenciennes, Hist. Nat. Poiss., XV, 1840, p. 284 (Cayenne); Kner, Sb. Ak. Wien, XVII, 1855, p. 128 (Para); Günther, Cat. Fish. Brit. Mus., V, 1864, p. 205 (copied); Eigenmann and Eigenmann, Proc. Cal. Acad. Sci., 2d Ser., I, 1888, p. 159; Occasional Papers Cal. Acad. Sci., 1890, p. 225 (Pará); Ribeiro, Peixes, IV, 1911, p. 212, Pl. XXXIX (Tabatinga; [1] Pará); Fisher, Ann. Carnegie Mus., XI, 1917, p. 419 (Pará).

[1] Ribeiro records from Tabatinga a specimen 810 mm. to the last scute, 960 mm. to the end of the caudal. This is the largest Doradid on record.

Doras papilionatus Filippi, in Guér. Ménev. Rev. and Mag. Zoöl., 1853, p. 167 (Amazons); Günther, l.c. (copied).

Doras lithogaster Heckel, MS. in Kner, Sb. Ak. Wien, XVII, 1855, p. 132 (Forte do Rio Branco); Günther, l.c. (copied).

Lithodoras lithogaster Bleeker, Nederl. Tijdsch. Dierk., I, 1863, p. 15 (name only).

Habitat: Para; Rio Negro; Cayenne.

Specimens Examined.

4248, I. U. M. one, 150 mm. Pará. Thayer Expedition?
4250, I. U. M. one, 171 mm. Brazil. Thayer Expedition.
16174, I. U. M. three, 128–164 mm. Belem, Pará, market. 1924. C. Ternetz.

Head 4.33; D. I, 6; A. 13 or 14; scutes 2 + 16 to 20.

Snout short, not greatly depressed; occipital region distinctly roof-shaped.

Eye large, lateral, 5.33 in head, 2 in snout, 2 in interorbital; the posterior margin of the pupil in the middle of the head. End of fontanel a little in advance of the posterior margin of the eye; a distinct groove from fontanel to the dorsal plate.

Gill-opening extending down to the middle of the pectoral spine. Isthmus almost equal to the length of the head; second scute about equal to the snout, 5th about six-tenths as long; two feeble plates in front of the hook-bearing ones on a straight line with the rest; caudal forked; mouth large, terminal; the lower jaw a little shorter than the upper, teeth in bands, strong; width in front of the pectoral a little greater than length of head; adipose equals anal, or shorter.

Genus IX. ACANTHODORAS Bleeker.

Type, *Silurus cataphractus* Linnæus.

Width at clavicle, 2.3 to 2.75 in the length; lateral plates numerous, covering more than half of the sides; no plates on upper or lower surface of caudal peduncle; adipose fin not continued forward as a keel; head depressed, the nuchal region but slightly arched; mouth terminal, the mouth wider than the interorbital; teeth well developed, in bands; no serræ on posterior margin of the dorsal spine; skull finely granular (the granules along its edge, larger, sharper in *spinosissimus*); fontanel subcircular, not continued as a groove; eyes very small, lateral, far in advance of the middle of the head; anterior nostril very near the lip; caudal rounded; opercle finely striate in *cataphractus*, coarsely granular in *spinosissimus;* preopercle and interopercle similar; preorbital short, in contact with the premaxillary in front, expanded, fan-like behind, with backward directed serræ on the outer half of the expanded edge; first suborbital small, along the lower surface of the nasal, second and third of about equal size (with sharp granules, similar to those along the edge of the skull above the eye in *spinosissimus*).

Coracoid entirely covered with skin or the tip finely granular.

First air-bladder short, heart-shaped; no trace of a second air-bladder.

Habitat: Amazon and Orinoco Basins; Guianas.

KEY TO THE SPECIES OF ACANTHODORAS.

a. A light line along the median spines of the lateral scutes and forward to, or near to, the eye; no plates on caudal peduncle above or below.

  b. Dorsal spine with a series of thorns along its anterior margin; lateral scutes 30, with a median hook only, covering one third of the side, not meeting above or below on the caudal peduncle; maxillary barbel reaching near tip of humeral process......................15. **calderonensis** (Vaillant).

  bb. Dorsal spine with series of thorns on the front and on the sides; lateral scutes covering more than half of the sides.

    c. Last four or five pairs of scutes separated by the narrow laminate caudal fulcra, the anterior plates of the caudal peduncle not quite meeting above and below; last scute narrow, with one hook; maxillary barbel to the tip of the humeral process; black, a light median line on top of head; a light line from posterior nostril along the edge of the granular part of the head and back along the hooks of the lateral scutes to the caudal; fins barred with black and white.

        16. **cataphractus** (Linnæus).

    cc. Lateral plates of the caudal peduncle meeting above and below, only the last one separated from its fellow by the caudal fulcra; last plate with several spines above and below the median hook; maxillary barbel reaching to the origin of the humeral process; black, a light line on top of head and a series of light spots along the middle of the back; lateral band similar to that of *cataphractus;* side of head and belly with conspicuous light spots; fins barred or spotted.

        17. **spinosissimus** (Eigenmann and Eigenmann).

15. Acanthodoras calderonensis (Vaillant).

Plate 10, Fig. 5.

Doras calderonensis Vaillant, Bull. Soc. Philom., Ser. 7, IV, 1880, p. 154 (Calderon); Eigenmann and Eigenmann, Occasional Papers, Cal. Acad. Sci., I, 1890, p. 234; Ribeiro, Peixes, IV, 1911, p. 206, Fig. 97.

Doras (Rhinodoras) depressus Steindachner, Flussf. Südam., II, 1881, 1, Pl. I, Figs. 3–3*a* (Lago Alexo).

Known from the types of *calderonensis* and *depressus*, the former in the Paris Museum, the latter, one specimen, 103 mm., from Lago Alexo, in the Vienna Museum.

16. Acanthodoras cataphractus (Linnæus).

Plate 4, Fig. 5; Plate 10, Figs. 1–4.

Silurus cataphractus Linnæus, Syst. Nat., ed. 10, I, 1758, p. 307; ed. 12, I, 1766, p. 506.

Doras cataphractus Cuvier and Valenciennes, Hist. Nat. Poiss., XV, 1840, p. 276 (?); Kner, Sb. Akad. Wiss. Wien, XVII, 1855, p. 126 (Rio Guaporé; Barra do Rio Negro); Bleeker, Ichth. Arch. Ind. Prodr., I, 1858, p. 54; Günther, Cat. Fish. Brit. Mus., V, 1864, p. 204 (?); Eigenmann and Eigenmann, Occasional Papers Cal. Acad. Sci., I, 1890, p. 234; Eigenmann, Repts. Princeton Univ. Exp. Patagonia, III, 1910, 393; Ribeiro, Peixes, IV, 1911, p. 208; Fisher, Ann. Carnegie Mus., XI, 1917, p. 419 (Maciél, Rio Guaporé).

Acanthodoras cataphractus Bleeker, Nederl. Tijdsch. Dierk., I, 1863, p. 17; Silures de Suriname, 1864, p. 40 (Surinam); Eigenmann, Mem. Carnegie Mus., V, 1912, p. 188 (Kangaruma; Georgetown; Gluck Island).

Cataphractus americanus Bloch and Schneider, Syst. Ichth., 1801, p. 107, Pl. 28; Lacépède, Hist. Nat. Poiss., V, 1803, p. 124, 127 (Carolina?).

Doras blochii Cuvier and Valenciennes, Hist. Nat. Poiss., XV, 1840, p. 277 (copied).

? *Doras castaneo-ventris* Schomburgk, Fishes Brit. Guiana, I, 1841, p. 161, Pl. 3 (Rio Pasawiri).

? *Doras brunnescens* Schomburgk, Fishes Brit. Guiana, I, 1841, p. 163 (Upper Essequibo).

Doras polyramma et *polygramma* (Ex Heckel, MS.) Kner, Sb. Akad. Wiss. Wien, XVII, 1855, pp. 126, 127.

Callichthys asper Gronow, Cat. Fish., ed. Gray, 1854, 157.

12032, I. U. M. 5, 58–88 mm. Gluck Island. Eigenmann.

Head 3.8 in the length; D. I, 5; A. 10 or 11; scutes 2 or 3 + 22 to 27. Eye 9 to 10 in head, 2.5 to 3 in snout, 3.5 to 4.5 in interorbital; fontanel subcircular, top of head scarcely arched; nuchal very little steeper; width at the clavicle equals distance from snout to dorsal plate, 2.6 to 2.75 in the length; mouth terminal; teeth numerous, bristle-like, in narrow bands.

Last scute of the caudal peduncle not extending over the entire depth of the peduncle, with a median spine only.

17. Acanthodoras spinosissimus (Eigenmann and Eigenmann).

Plate 2, Fig. 3; Plate 11, Figs. 1 to 4; Text-fig. 12.

Doras spinosissimus Eigenmann and Eigenmann, Proc. Cal. Acad. Sci., 2d Ser., 1888, p. 161; Occasional Papers Cal. Acad. Sci., I, 1890, p. 235 (Coary); Ribeiro, Peixes, IV, 1911, p. 209; Fisher, Ann. Carnegie Mus., XI, 1917, p. 420 (Maciél, Rio Guaporé). Ahl, Blätter Aquarien und Terrarienkunde 33, 1922, No. 1.

The type, a specimen 150 mm. long, in the Mus. Comp. Zoöl.

7167, C. M., 16005, I. U. M. 6, 50–111 mm. Maciél, Rio Guaporé. Haseman.

Head 3.6; D. I, 5; A. 11; scutes 21 to 23; eye 7.5 in head, 2.33 in snout, 3.5 in interorbital; skull very similar to *cataphractus* but with coarser granulations; jaws equal, mouth terminal, with well-developed teeth; width at clavicle 2.33–2.5 in the length; maxillary barbel about reaching pectoral; the last four scutes extending entirely across the caudal peduncle, with numerous spines; the last two separated above by the fulcra; the next two or three pairs in contact above and below; several pairs in front of the adipose in contact across the back.

Black, a light band along top of head; four light spots between dorsal and caudal along the back; a light band from upper edge of opercle to base of caudal; dorsal and pectoral with white spots or partial bars; ventral surface black, with irregular white spots; barbels all black and white banded.

Genus X. AGAMYXIS Cope.

Agamyxis Cope, Proc. Am. Philos. Soc., XVII, 1878, p. 101.

Type, *Agamyxis pectinifrons* Cope.

Eye moderate, entirely in anterior half of head, surrounded by serrate ridges; adipose fin not continued forward as a ridge; dorsal spine as long as pectoral spine, serrate in front and on the sides, smooth behind; lateral scutes covering entire side of caudal peduncle and more than upper half of area between anal and tip of humeral process; caudal peduncle with spiniferous plates above and below; nasal erect, pectinate; caudal rounded; breast covered with skin.

This genus differs from *Acanthodoras* largely in having the caudal peduncle covered with plates above and below.

KEY TO THE SPECIES OF AGAMYXIS.

a. Black, lower parts spotted with white, dorsal, anal, ventral and caudal with a yellowish cross bar; head 3.6; pectoral spine 3.2; eye 1.75 in snout, 6 in head; orbital margin raised, granulate, or pectinate; preopercular margin granular, the angle serrate; maxillary barbel not quite reaching base of pectoral spine. Scutes 27; caudal peduncle with spiniferous plates above and below.

                                                  18. **pectinifrons** [1] (Cope).

aa. Dark brown with irregular yellow spots scattered over entire body and fins; pectoral spine with cross bars; head 3.2; pectoral spine 3; eye 7.6 in the head, about 2.3 in the snout; orbital margin scarcely raised, prenasal bone moderately so; maxillary barbel extending beyond origin of pectoral; a blunt keel along the middle of the head, strongest on occipital region, fading out and dividing forward; teeth on the anterior edge of the dorsal stronger than those on the sides; 3 + 26 scutes, 4 to 5 spiniferous plates above and below on the caudal peduncle, continuous with the fulcra; types 90 and 110 mm.

                                               19. **flavopictus** (Steindachner).

18. **Agamyxis pectinifrons** (Cope).

Plate 16, Fig. 4.

Doras pectinifrons Cope, Proc. Am. Philos. Soc., XI, 1870, p. 568 (Pebas, Ecuador); Proc. Acad. Nat. Sci. Phila., 1871, Pl. 3; Eigenmann and Eigenmann, Occasional Papers Cal. Acad. Sci., I, 1890, p. 241 .

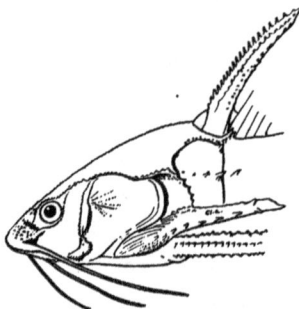

FIG. 19. *Agamyxis pectinifrons* Cope. The type in the collection of the Academy of Sciences, Philadelphia. Sketch of the type.

[1] In the original description Cope states "dorsal and pectoral spines elongate, both serrate before and behind."

Agamyxis pectinifrons Cope, Proc. Am. Philos. Soc., XVII, 1878, p. 679 (new genus).

Known from the type, 76 mm. long, in the collections of the Philadelphia Academy of Sciences.

19. Agamyxis flavopictus (Steindachner).

Doras (Agamyxis) flavopictus Steindachner, Sitzb. K. Akad. Wiss. Wien, 1908, No. VII, p. 3 (Iquitos).

Known from the types, two specimens 90 and 110 mm. long, from Iquitos in the Vienna Museum.

Genus XI. ASTRODORAS Bleeker.

Astrodoras Bleeker, Nederl. Tijdsch. Dierk., I, 1863, p. 17.

Type, *Doras asterifrons* Kner.

Width at clavicle 2.6 in the length; lateral plates numerous, covering half or more than half of the sides; caudal fulcra laminate, covering about half the caudal peduncle above and below; adipose fin not continued forward as a keel; head depressed, nuchal region roof-shaped; mouth terminal, equal to interorbital; teeth in bands; serræ on anterior margin of the dorsal spine, none on the sides or behind; skull finely granular or striate; fontanel elongate, continued as an obscure, interrupted groove; eye moderate, in center or a little in front of center of the head; anterior nostril near lip; caudal emarginate; preorbital short, in contact with the anterior *horn of the ethmoid* in front, its expanded posterior edge erect, with about 10 spines; suborbitals narrow, of equal width, the middle one granular. The anterior air-bladder heart- to kidney-shaped, merging into the posterior bladder without constriction. The latter forked to its base, the horns turned laterad, the tips slightly forward in a specimen No. 16003 I. U., from Jutahy; consisting of a short, minute, unforked diverticulum in two specimens No. 7163 C. M., from Santarem.

20. Astrodoras asterifrons (Heckel).

Plate 1, Fig. 11; Plate 4, Figs. 2 and 4.

Doras asterifrons Heckel MS., in Kner, Sb. Ak. Wiss. Wien., XVII, 1855, p. 123, Plate II, Fig. 2 (Barra do Rio Negro; Guaporé); Günther, Cat. Fish. Brit. Mus., V, 1864, p. 203 (Rio Cupai); Eigenmann and Eigenmann, Occasional Papers Cal. Acad. Sci., 2d Ser., I, 1890, p. 241 (Jutahy; Teffé; Porto do Moz; Serpa); Ribeiro, Peixes, 1911, p. 202, Fig. 96; Fisher, Ann. Carnegie Mus., XI, 1917, p. 420 (Maciél, Rio Guaporé; Santarem; San Joaquin).

Specimens Examined.

16003, I. U. M. one, 97 mm. to end of longest ray of upper caudal lobe. Jutahy. Thayer Expedition. 7163, C. M. 4, 67 to 89 mm. Santarem. Haseman.

Head 4; D. I, 6; A. 12; scutes 2 or 3 + 24 or 25, over half the depth; the rudimentary scutes on the tympanum above the line of the lateral hooks.

Eye 1.25 in snout, 5 in the head, 1.33 in the interorbital.

Width at clavicle slightly greater than distance from snout to posterior edge of dorsal plate; interorbital concave, the supercilian ridge raised; caudal fulcra laminate, the anterior ones above and below modified into a plate about the size of the eye, near the middle of the caudal peduncle; dorsal spine with two keels on each side; dorsal spine reaching nearly to or beyond origin of adipose; nasal erect, with about 12 spines; last three or four scutes with the median hook only; mouth terminal, teeth in narrow bands, the jaws equal; the abdominal area short; distance between tips of coracoids about 1.33 in distance from symphysis of coracoids to the anus; mottled ashen, darker above than below; fins spotted.

The difference in the air-bladders of the specimens from Jutahy and Santarem is so great that an unusually careful examination was made but yielded only the following differences:

a. Dorsal spine equal to its distance from the anterior edge of the eye; distance between the tips of the coracoids equals the distance between snout and end of dorsal plate; dorsal spine to middle of adipose. Second air-bladder consisting of two sausage-shaped tubes directed sideways. 16003, Jutahy specimen.

aa. Dorsal spine less than its distance from the anterior edge of the eye; distance between the tips of the coracoids equal to less than the distance from the symphysis of the coracoid to the base of the ventrals; greatest width a trifle less than the distance between the snout and the end of the dorsal plate; dorsal to or not quite to the adipose. Second air-bladder consisting of a short median protuberance. 7163 C. M., Santarem specimens.

Dorsal spine equals its distance from the nasal crest, reaching to the adipose according to the figure of the type.

Genus XII.   SCORPIODORAS Eigenmann, gen. nov.

Type, *Doras heckelii* Kner.

Having the characters of *Astrodoras*, except that the mouth is wider than the interorbital; nasal procumbent, the expanded edge raised but little; with more than 25 spines.

Anterior air-bladder kidney-shaped, large, its length about 1.5 in its width; posterior air-bladder banjo- or scorpion-shaped, the body of it heart-shaped, a posterior horn longer than the main part, recurved like the whip of a scorpion.

21. Scorpiodoras heckeli (Kner).

Plate 1, Fig. 30, Text-fig. 12, D.

Doras heckelii Kner, Sb. Akad. Wiss. Wien, XVII, 1855, p. 125, Fig. 4 (Rio Negro); Günther, Cat. Fish. Brit. Mus., V, 1864, p. 204; Eigenmann and Eigenmann, Occasional Papers Cal. Acad. Sci., I, 1890, p. 243 (Jutahy; Tonantins; Teffé; Tabatinga); Ribeiro, Peixes, IV, 1911, p. 205, Pl. XXXVIII, Fig. 2.

Habitat: Amazon basin.

16002, I. U. M. one, 161 mm. Teffé. Thayer Expedition.

Head 4.33; width at the clavicle 3.14 in the length, 1.25 in the distance between snout and edge of dorsal plate; D. I, 6; A. 11, scutes 2.5–3 in the depth, 4 + 29, those on the tympanum above the line of the hooks on the scutes; last five scutes without supplementary hooks; mouth terminal, the jaws equal; teeth numerous; nasal suberect, curved so that it seems continuous with the suborbital, with about 25 spines; the superciliary edge not raised; caudal fulcra laminate, the anterior ones not enlarged; dorsal spine with several ridges on the sides, not reaching adipose.

Abdominal area long, the distance between the tips of the coracoid processes 2 in the distance between the coracoid symphysis and the anus. Maxillary barbel extending past tip of humeral process.

Rust colored, the fins spotted; caudal with two longitudinal bands; ventral surface speckled.

Genus XIII. Amblydoras Bleeker.

Amblydoras Bleeker, Nederl. Tijdsch. Dierk., I, 1863, p. 86 (*affinis*).
Zathorax Cope, Proc. Acad. Nat. Sci. Phila., 1872, p. 271.

Type, *Doras affinis* Kner = *Doras hancocki* Cuv. and Val.

Width at clavicles 3 in the length; coracoid very much wider than clavicle, the surfaces of the two granular, the space for the ventral part of pectoral muscles reduced to a narrow tube in the anterior part of the thoracic buckler; scutes numerous, covering more than half the sides; caudal fulcra not covering half the caudal peduncle; adipose fin not continued forward as a keel; dorsal spine without hooks or thorns, grooved and ridged on sides and front; head depressed, the skull arched, finely striate-granular, without a median groove; eye in middle of head; anterior nostril near the lip; mouth terminal, the teeth well developed, in narrow bands; fontanel elongate; preorbital in contact with the premaxillary in front, its spines (about 15) covered; suborbitals narrow, of uniform width.

Anterior air-bladder in *hancocki* short, heart-shaped, without cœca. No posterior air-bladder.

Habitat: Guianas; Amazon.

KEY TO THE SPECIES OF AMBLYDORAS.

a. Maxillary barbel to posterior margin of orbit; pectoral spine reaching beyond ventrals; width at the clavicle 2.66 in the length; pale brown, the pectoral spine spotted with dark......22. **monitor** (Cope).

aa. Maxillary barbel reaching tip of humeral process; width at clavicle 3 in the length; pectoral spines reaching beyond base of ventrals; an irregular black band along sides below the median hooks. Three more or less distinct black blotches on the back, the last across the caudal peduncle.

23. **hancocki** (Cuvier and Valenciennes).

22. Amblydoras monitor (Cope).

Plate 16, Fig. 3; Text-fig. 20.

Zathorax monitor Cope, Proc. Acad. Nat. Sci. Phila., 1871, p. 271, Pl. 4, Fig. 1 (Amazon).

Doras monitor Eigenmann and Eigenmann, Occasional Papers Cal. Acad. Sci., I, 1890, p. 245.

Known only from the type in the collection of the Philadelphia Academy of Sciences.

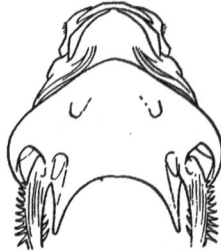

FIG. 20. *Amblydoras monitor* (Cope).   Sketch of thorax and lower side of head.

23. Amblydoras hancocki (Cuvier and Valenciennes).

Plate 3, Fig. 9; Plate 13, Figs. 1 to 4.

Doras costata Hancock, Zoöl. Journ., IV, p. 242 (Demerara).

Doras hancockii Cuvier and Valenciennes, Hist. Nat. Poiss., XV, 1840, p. 279; Günther, Cat. Fish. Brit. Mus., V, 1864, p. 202 (Demerara; Rio Cupai); Eigenmann, Mem. Carnegie Mus., V, 1912, p. 187 (Lama Stop-Off; Maduni Creek; Wismar; Rockstone; Gluck Island; Tumatumari; Rupununi Pan); Ribeiro, Peixes, IV, 1911, p. 210.

Doras affinis Kner, Sb. Ak. Wiss. Wien, XVII, 1855, 121, Pl. 11, Fig. 1 (Rio Branco; Guaporé); Günther, Cat. Fish. Brit. Mus., V, 1864, p. 202; Eigenmann and Eigenmann, Occasional Papers Cal. Acad. Sci., I, 1890, p. 238; Ribeiro, Peixes, IV, 1911, p. 201, Fig. 75.

Amblydoras affinis Bleeker, Nederl. Tijdsch. Dierk., I, 1863, p. 17.

Doras truncatus Bleeker, l.c., p. 18.

Habitat: Guiana to Guaporé.

Specimens Examined.

12033, I. U. M. 8, largest 110 mm.   Lama Stop-Off near Georgetown, Brit. Guiana.   Eigenmann.

12034, I. U. M. 2, 103 and 118 mm.   Maduni Creek near Georgetown.   Eigenmann.

12035, I. U. M. 3, about 37 mm.   Rockstone on the Essequibo, Brit. Guiana.   Eigenmann.

12036, I. U. M. one, 71 mm.   Tumatumari on the Potaro, Brit. Guiana.   Eigenmann.

12037, I. U. M. 2, 36 and 40 mm.   Rupununi Pan, Brit. Guiana.   Eigenmann.

15887, I. U. M. one, 65 mm.   Itaya above Iquitos, Peru.   Allen.

17040, I. U. M. 17, 60–98 mm.   Lagoon near Reyes, Bolivia, Oct. 1921.   Pearson.

Head 4 to 4.3; D. I, 6; A. 11 to 13; eye 4.5 in the head, 1.5 in the snout, 1.5 in the interorbital; width of the clavicle 3 in the length.

Maxillary barbel about to tip of humeral process; scutes about half as deep as the body; caudal fulcra laminate, none converted into a plate; caudal slightly emarginate; a light line along the scutes; a black stripe from humeral process to end of anal below the scutes; a black band across caudal behind the last scute, a number of saddle-shaped spots on middle of back; upper parts dark ashy with a few spots smaller than pupil; spots on dorsal and caudal.

Genus XIV. ANADORAS Eigenmann, gen. nov.

Type, *Doras grypus* Cope.

Width between clavicles less than 3 in the length; scutes numerous, covering less than one fourth of the sides; modified fulcra covering much of the caudal peduncle above and below.

Adipose fin short; not continued as a keel; dorsal spine grooved; without serræ; no plates between the dorsal and adipose; granulations on breast inconspicuous, narrow, or confined to the tip of the coracoid. Eye entirely in anterior part of the head; preorbital bone inconspicuous, snout depressed, mouth terminal; teeth minute, in bands; fontanel oval, not continued as a groove; head minutely granular.

First air-bladder heart-shaped, with a median, longitudinal groove, without marginal prolongations; second air-bladder reduced to a minute tubule (*weddellii*) or entirely suppressed (*regani, grypus*) (5, Text-fig. 2).

Habitat: Amazon Basin.

KEY TO THE SPECIES OF ANADORAS.

a. Sides with a pale lateral band.

    b. Dorsal region copper-colored, with few faint or sepia spots; lateral plates yellow, the region below brownish yellow; caudal with dusky stripes along the upper and lower lobes; lateral plates 3 + 26-27, narrow, leaving wide naked areas above and below. Eye small, 6 in head, 2 in interorbital, 1.5 in snout; interorbital flat; maxillary barbel extending beyond base of pectoral; each lateral plate with one or two points above and below the median hook. Caudal peduncle with 7-10 bony plates, grading into the fulcra above and below; caudal emarginate; caudal lobes with violet bands. 90 mm.

                                            24. regani (Steindachner).

    bb. Three or more dark blotches, below the band, on a light background; the dark above the band with darker, light-edged, blotches; anterior parts of dorsal, anal and ventrals black, the posterior part light, spotted; middle and edges of caudal light, the base and a stripe along the lobes black; coracoid process flaring backward; distance from union of coracoids to the line uniting the tips one half to one third the distance between the tips; caudal deeply emarginate...........**25. grypus** (Cope).

    bbb. Brown above, with a yellow lateral band; caudal with vertical dark bars. Orbit 1 in interorbital, 3 in head; maxillary barbels reaching end of pectoral spines; caudal truncate, scutes 26, leaving a broad naked band above and below, with only a single median spine. 82 mm.

                                            26. nauticus (Cope).

22

aa. Sides without a lateral stripe.

   c. Irregularly spotted; maxillary barbel reaching pectoral spine; opercle granular; eye 3 to 4 in inter-
      orbital; nasal without a serrated crest; fontanel oval, with a granular border; coracoid processes
      flaring but little; the distance between the juncture of the coracoids and a line connecting their
      tips ⅘ to ⅗ of the distance between the tips; outer face of striate portion of the coracoid convex;
      scutes 25–27; the caudal fulcra more plate-like than in *grypus*, covering most of the caudal
      peduncle...27. **weddellii** (Castelnau).

   cc. Yellowish with sepia markings, 4 bars on dorsal spine, 5 on pectoral, 4 on posterior part of body;
      maxillary barbel reaching tip of humeral process; opercle striate; nasal strongly pectinate;
      opercle reticulate; eye 1.75 in interorbital; scutes 29, with an auxiliary spine above and below;
      top of head striate; coracoid process reaching middle of pectoral spine; the humeral process
      beyond the third quarter of that spine; caudal peduncle with six plates...28. **insculptus** (Ribeiro).

24. **Anadoras regani** (Steindachner).

Text-fig. 21.

Doras regani Steindachner, Sitzb. K. Akad. Wiss. Wien, 1908, Nr. XI (Pará).

Known from four specimens, 90 mm. long, in the Vienna Museum, and 16191,
I. U. M. 2, 123 and 140 mm.   Santarem market.   Sept. 1924.   C. Ternetz.

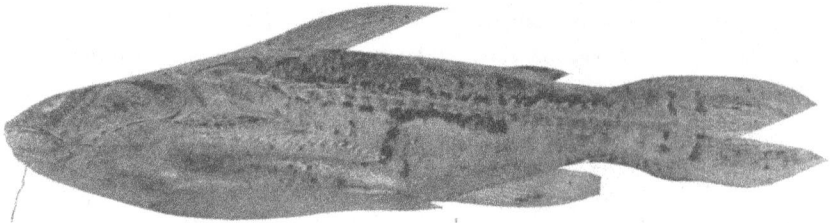

Fig. 21. *Anadoras regani* (Steindachner).   140 mm.   Photo by N. E. Pearson.

25. **Anadoras grypus** (Cope).

Plate 4, Fig. 7; Plate 15, Figs. 1 to 3.

Doras grypus Cope, Proc. Acad. Nat. Sci. Phila., 1871, p. 270, Pl. XV, Figs. 1–1a (Ambyiacu River).
Doras weddellii Eigenmann and Eigenmann, Occasional Papers Cal. Acad. Sci., I, 1890, p. 239 (in part).

Specimens Examined.

4249, I. U. M. one, 129 mm.   Teffé.   Thayer Expedition.
15849, I. U. M. 50 to 122 mm. to end of middle caudal rays.   Lake Cashiboya.   Allen.
15961, I. U. M. 3, 80 to about 150 mm.   Iquitos.   Morris.

Head nearly 4 in the length; D. I, 6; A. about 12; width at clavicle nearly 3 in
the length; eye 5.6 in the head, 1.75 in snout, 2.75 in interorbital; maxillary barbel
reaching beyond base of pectoral spine; mouth terminal, the teeth in narrow bands;

scutes 26 or 27, narrow, about 3 in the depth; rudimentary scutes feeble, concealed; laminate caudal fulcra covering about half the caudal peduncle above and below; nasal bone not serrate; skull finely granular. For color see the key.

26. Anadoras nauticus (Cope).

Zathorax nauticus Cope, Proc. Acad. Nat. Sci. Phila., 1874, p. 133 (Nauta); Proc. Am. Philos. Soc., XVII, 1878, p. 678; Eigenmann and Eigenmann, Occasional Papers Cal. Acad. Sci., I, 1890, p. 246.

Known from the type, 82 mm. long, in the Philadelphia Academy of Natural Sciences.

27. Anadoras weddellii (Castelnau).

Plate 4, Fig. 6; Plate 15, Fig. 4.

Doras weddellii Castelnau, Anim. Amer. Sud., 1855, p. 48, Pl. XVII, Fig. 1 (Chiquitos); Günther, Cat. Fish. Brit. Mus., V, 1864, p. 203 (Santarem); Vaillant, Bull. Soc. Philom., Ser. 7, IV, 1880, p. 154 (Calderon); Eigenmann and Eigenmann, Proc. Cal. Acad. Sci., 2d Ser., I, 1888, 163 (in part); Occasional Papers Cal. Acad. Sci., I, 1890, p. 239 (the specimens with a restricted coracoidal striation) (Fonteboa; Teffé; Serpa; Porto do Moz; Silva, L. Saraca); Ribeiro, Peixes, IV, 1911, p. 202; Fisher, Ann. Carnegie Mus., XI, 1917 (Manaos; Santarem; Rio Jauru; Bastos, Rio Alegre and Maciél into Guaporé; San Joaquin, Bolivia).

Specimens Examined.

5110, I. U. M. 2, 112 and 113, the smaller a skeleton. Island Marajo. Thayer Exp.
10218, I. U. M. one, about 75 mm. to end of middle caudal rays. Rio Pilcomayo.

Head 3.66; clavicular width about 3 in the length; eye 7 in the head, 2 in the snout, 3 in the interorbital. D. I, 6; A. 12; scutes 3 + 26 or 27, one fourth or one fifth of the depth of the body; caudal fulcra heavy, laminate, reaching to the adipose and nearly to the anal; mouth terminal; teeth well developed, in bands; nasal not serrate.

28. Anadoras insculptus (Ribeiro).

Plate 14, Figs. 1 to 3.

Doras insculptus Ribeiro, Comm. Linhas. Telegr. Estrat. Matto Grosso as Amazonas, Annexo 5, 1912, p. 22 (Manaos).

Known from the types, the largest 100 mm., in the Mus. Nacional Rio de Janeiro. I am indebted for the figures to Dr. A. Ribeiro.

Genus XV. HYPODORAS Eigenmann, gen. nov.

Type, *Hypodoras forficulatus* Eigenmann.

Adipose fin but little shorter than the anal fin, the anterior two thirds covered by a rhomboidal plate; caudal fulcra laminate, covering the entire caudal peduncle above and below; dorsal spine striate in front and on the side, without serræ; lateral scutes of the caudal peduncle in contact with the plates above and below, with a single hook; those in front of the caudal peduncle with 2 to 5 hooks above the median one and with 1 to 2 below it; humeral process very strong, with a series of spines; process of coracoid striate; eye immediately in front of the middle of the head; preorbital crest pectinate; suborbital granular.

Anterior air-bladder (Fig. 4, Plate 2) wider than long, kidney-shaped, without marginal tentacles or cœca, with a median longitudinal constriction, directly continued into the posterior air-bladder, which, at the point of union, is about one third as wide as the anterior and forked (divided into two horns) in its posterior three fifths. The horns seem to be continuations of the two lateral lobes of the anterior bladder. One of the horns is turned sidewards.

This genus differs from all the others of the Doradidæ in the plate over the anterior part of the adipose and in the nature of its air-bladder.

29. Hypodoras forficulatus Eigenmann, spec. nov.
Plate 4, Fig. 1; Plate 25, Fig. 3; Text-fig. 12, I.

15876, I. U. M. one, *type*, 123 mm., 104 mm. to end of last scute.   Iquitos.   Allen.   Sept. 1920.

This species has the general appearance of *Anadoras grypus* Cope but differs from *grypus* very notably.   The data in parentheses refer to a specimen of *D. grypus*, 108 mm. long to the end of the last scute.

Head 3.8; depth 5.3; D. I, 6; A. 11; scutes 2 + 26.   Top of head granulose, the fontanel not continued as a groove; prenasal bone deeply pectinate (smooth); humeral region very wide; equal to the distance from snout to dorsal fulcrum; interorbital equals snout (snout and eye); eye 2.66 in interorbital, 6 in the head; teeth minute, in bands; maxillary barbel extending a little beyond coracoid process, banded; outer mental barbel reaching coracoid, inner mental barbel considerably shorter; dorsal spine deeply striate in front and on the sides, smooth behind, a little over half as long as the heavy, serrate, pectoral spine, which reaches the middle of the ventrals; ventrals broad, rounded; anal rounded; caudal truncate; dorsal and pectorals lanceolate; humeral process very strong, reaching to below 4th dorsal ray, with teeth toward its tip equal to those of the pectoral which oppose them; coracoid process striate, narrow, reaching about to third fifth of the pectoral spine; caudal peduncle covered with plates below (naked); a single large plate between the adipose and the numerous caudal fulcra (none); a very large keeled plate over anterior part of the adipose (none); scutes very high (low), covering the entire sides, two or three from the dorsal plate to the humeral process, those of the sides crowded; the median spine moderate, with up to 5 smaller spines above the larger one and two, rarely three, below the main spine in the region of the anal (none); one or none on the caudal peduncle above and

·below the central spine; the central spines of nearly uniform size from below the dorsal to the caudal; ashen, a narrow black bar from posterior half of dorsal down and back to a black spot at the end of the humeral process, then downward and forward to the edge of the belly; a wider bar downward and forward from the soft part of the adipose, then to and partly across the base of the anal; a bar across the middle of the caudal peduncle and another across the base of the caudal; dorsal black; the distal part hyaline, spotted with black; caudal, anal, ventrals and pectorals hyaline, with black spots, bases of caudal, anal and ventrals with a larger black spot; ventral surface dark, clouded.

Genus XVI. PHYSOPYXIS Cope.

Physopyxis Cope, Proc. Acad. Nat. Sci. Phila., 1871, pp. 112, 272, 273.

Type, *Physopyxis lyra* Cope.

Eye small, in anterior half of head; no adipose; dorsal spine serrate on lower half of anterior margin; pectoral spine extending beyond origin of anal; caudal truncate; clavicles and coracoid granular, developed into a large armor.

30. **Physopyxis lyra** Cope.

Plate 16, Figs. 1 and 2.   Text-fig. 22.

Physopyxis lyra Cope, l.c. (Ambyiacu River).

This species is known from the type 35 mm. long in the collections of the Philadelphia Academy of Sciences.

FIG. 22.   Dorsal spine of the type of *Physopyxis lyra*.

Genus XVII. PSEUDODORAS Bleeker.

Pseudodoras Bleeker, Ichthyol. Arch. Indici Siluri, 1858, p. 53 (to include *Doras niger* Valenciennes; *D. brevis* Kner, *D. humeralis* Kner; *D. lipophthalmus* K.).

Oxydoras Bleeker, Nederl. Tijdsch. Dierk., I, 1863, p. 14 (type, *Doras niger* Val. = *Pseudodoras niger* Bleeker).

Type, *Doras niger* Valenciennes.

Bleeker in 1858 based his *Pseudodoras* on *Doras niger* Val., *D. brevis* Kner, *D. humeralis* Kner and *D. lipophthalmus* Kner, in the order named.   This is a miscellaneous collection of fishes, evidently known to Bleeker only as a series of names. There is no evidence that Bleeker considered the *Doras niger* as anything except Valenciennes' species which should remain the type of the name *Pseudodoras*. If there is any doubt on this score it is removed by Bleeker (1863, p. 14) when he distinctly states that his *Pseudodoras niger* is the *Doras niger* of Val. and not the *Doras niger* of Kner.

In Nederl. Tijdsch. Dierk., I, 1863, pp. 12, 14, he endeavored to suppress his *Pseudodoras:* "De plus ma première espèce de *Pseudodoras* est le *Doras niger*, pour lequel le nom d'*Oxydoras* a droit de priorité, et les trois autre espèces, ainsi que le *Hemidoras stenopeltis* sont de vrais Doras, tel que je comprends actuellement le genre Lacépèdien, dont en effet le *Doras carinatus* espèce voisine, est le type."

Bleeker, however, is wrong in assuming that the name *Oxydoras* of Kner-has priority to the name *Doras niger* Valenciennes. Kner did not know *Doras niger* of Valenciennes when he framed his name *Oxydoras*, he simply mistook specimens of another species for it. The *Doras niger* Val. cannot be made the type of *Oxydoras*. *Doras niger* of Kner is not the *Doras niger* of Valenciennes, a fact Bleeker had discovered by 1863. On p. 14, 1863, Bleeker makes *Doras niger* Val.—a species not known to Kner—the type of Kner's *Oxydoras*, but he reconsidered this and on page 85, he distinctly makes *Doras (Oxydoras) niger* Kner (= *Oxydoras kneri* Bleeker) the type of *Oxydoras*. As this was the first and a perfectly legitimate restriction of the genus *Oxydoras*, *O. kneri* should remain the type of *Oxydoras*. Since the *Doras niger* of Kner belongs to a genus distinct from the *Doras niger* of Valenciennes there is no reason why the name *Pseudodoras*, with *Doras niger* of Valenciennes as its type, should be suppressed and it is herewith resurrected.

Width at clavicles less than length of head; coracoid covered with skin; lateral plates few, narrow, 3 + 18 to 23; adipose fin continued forward as a keel; head not depressed, snout conical; mouth inferior; premaxillaries rhomboidal; teeth feeble; serration on posterior margin of dorsal spine weaker than on the anterior; occipital region roof-shaped, granular or striate, without a median groove; eyes lateral, entirely behind the middle of the head; anterior nostril remote from the lips; fontanel double; caudal emarginate; opercle striate; suborbital granular in old. Several tentacles across the roof of the mouth between the second and third pairs of gill-arches, two rows of tentacles converging from in front of the first arch forward, ending in a ridge; smaller ones on the floor of mouth opposite; preorbital comma-shaped, the front incurved, connected with the ethmoid. First suborbital in part under the preorbital, the second large, only half of it forming part of the orbital border, the third very slender (3, Plate 11).

Air-bladders of *P. niger* (Fig. 4, Plate 2) both well developed, with marginal indications of lobules or beginnings of cœca. Anterior bladder elongate, heart-shaped, its width 1.25 in its length. Posterior bladder elliptical, a little more than twice as long as wide; its length less than 2 in the length of the anterior.

The generic description is based on *P. niger*.

KEY TO THE SPECIES OF PSEUDODORAS.

a. Gill-opening extending far below pectoral spine; dorsal spine with feeble teeth on the distal part of the posterior margin, disappearing with age; no conspicuous caudal fulcra; occipital region steeply roof-shaped; eye in posterior half of head, 3–4 in interorbital; highest scutes 1.3–1.5 in interorbital; three small bones in front of the hook-bearing ones; width in front of pectoral equals head less one fifth to three fifths of opercle. (A small granular plate between the upper halves of the first two hook-bearing lateral scutes in the old.)..31. **niger** (Valenciennes).

aa. Gill-openings extending to the base of the pectoral spine; dorsal spine smooth behind (?); caudal fulcra extending halfway to anal and adipose; occipital region obtusely roof-shaped; eye just behind the middle of the head, 3 in the interorbital; scutes 16, high; two small bones in front of the spine-bearing scutes. A golden stripe along middle line from dorsal to end of adipose, an ill-defined dark stripe along the middle of the head; lateral parts of top of head, middle of opercle, lower part of cheeks and lateral plates golden...32. **huberi** (Steindachner).

31. **Pseudodoras niger** (Valenciennes).

Plate 1, Fig. 16; Plate 17, Figs. 1–4; Plate 23, Fig. 1; Text-figs. 2, 6, 7 and 10.

Doras niger Valenciennes, in Humboldt, Rec. d'Observ. Zoöl. et Anat. Comp., II, 1811, p. 184; Cuvier and Valenciennes, Hist. Nat. Poiss., XV, 1840, p. 291 (?); Schomburgk, Fishes Brit. Guiana, I, 1841, p. 165; Müller and Troschel, in Schomburgk, Reisen, III, 1848, p. 629 (Rivers of Guiana); Bleeker, Nederl. Tijdsch. Dierk., I, 1863, p. 14 (name only); Eigenmann, Mem. Carnegie Mus., V, 1912, p. 190 (Rupununi).

Rhinodoras niger Günther, Cat. Fish. Brit. Mus., V, 1864, p. 209 (Amazons); Cope, Proc. Am. Philos. Soc., XVII, 1878, 678 (Nauta); Vaillant, Bull. Soc. Philom. (7), IV, 1880, p. 14 (Calderon).

Oxydoras niger Eigenmann and Eigenmann, Proc. Cal. Acad. Sci. (2), I, 1888, p. 159 (Teffé; Gurupa; Manacapuru; Coary; Obidos); Occasional Papers Cal. Acad. Sci., I, 1890, p. 247; Kindle, Ann. N. Y. Acad. Sci., VIII, 1894, p. 251 (Pará); Eigenmann, Repts. Princeton Univ. Exp. Patagonia, III, 1910, p. 393; Ribeiro, Peixes, IV, 1911, p. 193, Pl. XXXVII; Fisher, Ann. Carnegie Mus., XI, 1917, p. 420 (Santarem; Manaos).

Doras humboldti Agassiz, Selecta Gen. et Spec. Pisc. Bras., 1829, p. 129, Pl. 5 (Rio San Francisco, Brazil); Agassiz, A Journey in Brazil, 1868, ——.

Corydoras edentatus Spix, Selecta Gen. et Spec. Pisc. Bras., 1829, Pl. 5.

Rhinodoras prionomus Cope, Proc. Acad. Nat. Sci. Phila., 1874, p. 134 (Nauta); Proc. Am. Philos. Soc., XVII, 1878, 678 (Nauta).

Rhinodoras teffeanus Steindachner, Sb. Akad. Wiss. Wien, LXXI, 1875, 145, Pl. 3 (Teffé).

This species is known from near sea level to about 3000 feet. In British Guiana, the Amazons from Pará westward to the Marañon, from the Rio San Francisco, and from the following specimens in the collections of Indiana University:

5105, I. U. M. one, 234 mm. to end of middle caudal rays. Pará. Thayer Exp.

15651, I. U. M. 15960, 4, 170–260 mm. Iquitos. Allen, 1920, Morris, 1922.

15450, I. U. M. one, about 250 mm. Yarinacocha. Allen. Sept. 1920.

15845, I. U. M. one, about 200 mm. Lake Cashiboya. Allen. 1920.

15850, I. U. M. one, skin, 202 mm. from snout to end of dorsal plate, along the median line; about 650 mm.
over all. La Merced, Peru, 2500 feet. Sr. Ecuador Præli. 1918.

16004, I. U. M. one, 273 mm. to end of the scutes. Rio Pacaya. Allen.

   743, T. R. S. one, skin, 965 mm. to end of caudal, Kartabo, Guiana. Beebe.[1]

16169, I. U. M. one, 453 mm. Santarem. 1924. Carl Ternetz.

LOCALITY: Kartabo, Bartica District, British Guiana. Date: 24, VIII, 1922.

Dimensions: From field measurements.   In mm.

Length	840
Head	245
Depth (in front of dorsal fin)	169
Greatest width (anterior to pectoral fins)	175
Depth of caudal peduncle	41
Width of caudal peduncle	53
Eye socket	25.2 x 16.5
Interorbital	77
Eye to posterior nostril	54
Asymmetrical:	
Right	32
Left	21
Interpostnasal	54.5
Dorsal spine	166
Pectoral spine	177
Width of mouth	88
Snout to dorsal	320
Snout to anal	570
Snout to pectoral	210
(The asymmetry in the following four measurements is due to accidental injury.)	
Right maxillary barbel	94
Left maxillary barbel	115
Left inner mental barbel	36
Left outer mental barbel	55
Pectoral	I, 9
Dorsal	I, 6
Ventrals	7
Anal	9
Weight	16 pounds.

Brownish black above, dirty white below. Fins and mouth black. Lateral scutes yellowish horn. Spines ivory white.

FOOD: Flower buds, petals, seeds.

EYE: Sunk beneath surface of head. Iris 15 mm. in diameter. Purplish brown, flecked with yellowish white, the coloration most prominent anteriorly and posteriorly. A narrow pupil rim, yellow, widest posteriorly, and sending downward on lower side of the eye a single irregular line of whitish yellow, passing halfway through the iris. The Creoles call this fish "Fork-snout," and claim that it is an adult *Hemidoras carinatus*, which is also known by that name. This specimen was caught in a drift net in square B7 at Kartabo.

[1] The following notes accompanied the large specimen taken by Dr. W. Beebe.
Specimen Number: T.R.S. 743. Skin only.

D. I, 6; A. 11, scutes 3 + 18 to 23; head 3 to 3.75 in the length to the last scute; eye 8 in the head, 4.5 in the snout, 3 in the interorbital in smallest, respectively 10.5, 5.5, 3.33 in the largest; from snout to dorsal fulcrum 2.4–2.66 in the length to end of scutes. Bones of head striate in young and half grown, becoming striate granular; premaxillaries subrhomboidal, about the size of eye. Teeth very feeble or none.

Dark brown, the fins black.

In the specimen from La Merced (Text-fig. 10), a small plate, similar to the tympanic plates is interpolated between the first two hook-bearing ones and in a line with the distant rudimentary ones. Scutes 3 + 1 + 1 + 17 or 18 (22 in the largest).

32. Pseudodoras huberi (Steindachner).

Oxydoras (Rhinodoras) huberi Steindachner, Sitzb. K. Akad. Wiss. Wien, 1911, CXX, p. 324 (Rio Tocantins at Cameta).

This species is known from the type, 400 mm. long, taken at Cameta by Dr. Snethlage. It is now in the Vienna Museum.

Head 3.4; D. I, 6; A. 12; scutes 2 + 17; eye $8\frac{2}{11}$ in the head, 4 in snout, about 3 in interorbital. Bones of head striate granular; bristle-like teeth in moderate bands; maxillary barbel to tip of humeral process; posterior nostril equidistant from the anterior and the eye; pectoral spine longer than the head.

Black, with a lighter line along the middle of the back, from dorsal to end of adipose, and brown to yellow markings on head and sides.

Genus XVIII. Oxydoras Kner.

Oxydoras Kner, Sb. Ak. Wiss. Wien, XVII, 1855 (various species: *D. stenopeltis* Kner; *carinatus* C. and V.; *niger* Kner, *non* C. and V.; *lipophthalmus* Kner; *d'Orbignyi* Kröyer).
Oxydoras Bleeker, Nederl. Tijdsch. Dierk., I, 1863, p. 85 (reprint, p. 9).

Type, *Oxydoras kneri* Bleeker, based on *Doras (Oxydoras) niger* Kner, *non Doras niger* Valenciennes.

The status of the name *Oxydoras* as restricted by Bleeker is given under *Pseudodoras.*

In 1863 Bleeker, p. 14, restricted the genus *Oxydoras* of Kner to specimens supposedly belonging to *Doras niger* Valenciennes. But Kner's specimens did not belong to *Doras niger* Val.

Later in the same year, p. 85, Bleeker restricted it by selecting as type the species named *niger* by Kner but which was, in reality, a distinct genus and species named *kneri* by Bleeker.

Similar to *Pseudodoras*, the number of lateral plates much larger, no tentacles on roof of mouth.

Habitat: Paraguay Basin.

33. Oxydoras kneri Bleeker.
Plate 1, Fig. 15.

Doras (Oxydoras) niger Kner (*non Doras niger* Val.), Sb. Ak. Wiss. Wien, XVII, 1855, p. 146 (Cujaba).

Oxydoras kneri Bleeker, Nederl. Tijdsch. Dierk., I, pp. 12 and 14 (substitute name for *Doras niger* Kner *non* Valenciennes); Eigenmann, Proc. Acad. Nat. Sci. Phila., 1903, p. 501 (Asuncion); Ribeiro, Peixes, IV, 1911, p. 194; Fisher, Ann. Carnegie Mus., XI, 1917, p. 420 (Corumba).

Rhinodoras knerii Günther, Cat. Fish. Brit. Mus., V, 1864, p. 209.

The type, 15 inches long, is in the Vienna Museum. It has 34 lateral scutes, those on the caudal peduncle deeper.

9840, one, 205 mm. to end of scutes. Asuncion. Ternetz.

Head 3.5; D. I, 6; A. 12; scutes 3 + 34–40; eye 10 in the length of the head, 5.5 in the snout, 3 in interorbital; distance from tip of snout to dorsal fulcrum 2.66 in the length to end of scutes; top of head and opercle striate-granular.

Genus XIX. RHINODORAS Bleeker.

Rhinodoras Bleeker, Nederl. Tijdsch. Dierk., I, 1863, p. 14.

Type, *Doras (Oxydoras) d'Orbignyi* Kröyer MS.

Eye medium, in middle of head; adipose fin prolonged forward as a keel; dorsal spine strongly serrate, the serræ stronger on posterior margin; lateral scutes low, about 30; caudal forked; caudal peduncle covered with plates (modified fulcra) above and below; preorbital plates obscure; head subconical; fontanel continued as an obscure groove to the dorsal. Air-bladders (Fig. 8, Plate 2) similar to those of *Pseudodoras niger* but without indications of cœca, the anterior bladder less depressed; fontanel not divided, not continued as a groove. Mental barbels in two distinct pairs.

34. Rhinodoras d'orbignyi (Kröyer) MS.
Plate 1, Fig. 18; Plate 26, Fig. 5.

Doras (Oxydoras) d'orbignyi (Kröyer) Kner, Sb. Ak. Wiss. Wien, XVII, 1855, p. 149, Pl. V, Fig. 9 (Rio Plata).

Doras d'Orbignii Hyrtle, Denk. Ak. Wiss. Wien, XVI, 1859, p. 17 (vertebræ 6 + 4 + 21).

Rhinodoras orbignyi Bleeker, Nederl. Tijdsch. Dierk., I, 1863, p. 84 (name); Günther, Cat. Fish. Brit. Mus., V, 1864, p. 209 (copied).

Oxydoras d'orbignyi Eigenmann and Eigenmann, Proc. Cal. Acad. Sci., 2d Ser., I, 1888, p. 159; Occasional Papers Cal. Acad. Sci., I, 1890, p. 249.

Doras nebulosus Eigenmann and Kennedy, Proc. Acad. Nat. Sci. Phila., July 1903, p. 500 (Asuncion); Ribeiro, Peixes, IV, 1911, p. 213.

9837, I. U. M. one, 160 mm. (154 mm. at present), probably Asuncion. Carlos Ternetz.

Head 3.66; D. I, 6; A. 13; scutes 1 + 29 or 30, width at clavicle 1.15 in head, about 4 in the length; eye 7.5 in the head, 3.5 in snout, 2 in interorbital; posterior nostril equidistant from anterior and lip; bones of head obscurely striate or granular; mouth inferior, teeth strong, in short bands; maxillary barbel not reaching gill-opening.

Sides and fins sharply clouded.

Genus XX. TRACHYDORAS Eigenmann, gen. nov.

Type, *Trachydoras atripes* Eigenmann.

Snout, preopercle and opercle granular; short and deep; coracoid process partly or wholly granular; first lateral scute very large, connecting the humeral process and the descending process of the dorsal plate; humeral process long and narrow, more or less obliquely truncate; fontanel not continued as a groove; no foramen in the dorsal buckler; snout short, anterior nostrils equidistant from tip of snout and the posterior nostrils or nearer the former; mouth very small; teeth feeble, or none; maxillary barbels just reaching gill-opening or shorter; adipose fin about half the length of the anal, sometimes continued forward as a ridge; depth 3–4.5; one or more pectoral pores.

Habitat: Paraguay and Amazon Basins.

KEY TO THE SPECIES OF TRACHYDORAS.[1]

a. Interopercle covered with skin; hooks of the lateral scutes along the line of the highest point in upper margin of the humeral process; first scute not reaching the lower margin of the humeral process, posterior margin of the humeral process obliquely emarginate.

  b. Coracoid process much shorter than the humeral process, granular on part of its length; opercle, suborbital line and preopercular line granular; depth 4 in the length to the end of lateral hooks, equal to the length of the head; dorsal spine equal to head or head and width of humeral granulation; pectoral spine longer; depth of caudal peduncle 4 in the head; eye 1.25 in interorbital, 1.66 in snout, lateral plates .33 in the depth; posterior nostril nearer anterior nostril than eye; dorsal fulcrum dark, middle caudal rays hyaline.............................35. **nattereri** (Steindachner).

  bb. Coracoid process a trifle longer than the humeral process, striate its entire length; suborbital line covered with skin; opercle and preopercular line striate; depth 3.2 in the length, greater than head and humeral granulations; dorsal spine equals length of head; pectoral spine considerably longer; depth of caudal peduncle 3 in the head; eye 1.4 in interorbital, 3.1 in the head; lateral plates .33 of the depth; posterior nostril nearer eye than to anterior nostril; base of middle caudal rays dusky...36. **paraguayensis** (Eigenmann and Ward).

aa. Interopercle granular; hooks of the scutes above the line of the upper margin of humeral process; first scute crescent-shaped, extending to the lower margin of the humeral process.

  c. Dorsal spine equals head or a little shorter than head, about equal to the pectoral spine; depth of caudal peduncle about 3.8 in the head; eye 1–1.25 in interorbital, equal to snout behind the anterior nostrils; depth of lateral plates about 2.3 to 2.6 in the depth; posterior nostril much nearer eye than to anterior nostril; scutes 35; maxillary barbel with many barblets........37. **atripes** Eigenmann.

[1] *D. microstomus,* of which only small specimens, 52 mm. and shorter, have been taken, may belong here; in general appearance, mouth, barbels, dorsal buckler, etc., it resembles the species of this group. The preopercle, opercle, etc., are not granular, but the preopercular edge extends as a ridge across the cheek. *D. brevis* also resembles this group.

cc. Dorsal spine equals head and a little more than width of the humeral granulation, about equal to the pectoral spine; depth of lateral plates about half the depth; posterior nostril much nearer eye than to anterior nostrils; D. I, 6; A. 13; scutes 33–34. Maxillary barbel with but one barblet; all fins hyaline; eye equals interorbital.............................38. **trachyparia** (Boulenger).

35. **Trachydoras nattereri** (Steindachner).

Plate 18, Figs. 1 and 2; Plate 26, Figs. 2 and 3.

Oxydoras nattereri Steindachner, Flussf. Südam., II, 1881, p. 4, Pl. 2, Figs. 1–1*a*. (Teffé.)

Hemidoras nattereri Eigenmann and Eigenmann, Proc. Cal. Acad. Sci. (2), I, 1888, p. 158 (Coary; Jutahy; Teffé); Occasional Papers Cal. Acad. Sci., I, 1890, p. 253; Fisher (part), Ann. Carnegie Mus., XI, 1917, p. 421 (San Joaquin, Bolivia; Santarem; Maciél; Rio Guaporé).

15967, I. U. M. one, 130 mm. Iquitos. Morris. 1922.

Head 3.6; depth 3.5; D. I, 6; A. 14; scutes 3 + 32; eye 1.5 in snout, 3.5 in head, 1.25 in interorbital.

Snout blunt, occipital region roof-shaped; humeral process more finely striate than in *T. paraguayensis*, the striation not more coarse toward the upper margin; clavicular and coracoid granulation separated in front of the pectoral by an area covered with skin; coracoid process reaching to middle of humeral process; maxillary barbel with about 7 barblets; rudimentary scutes without a hook. Line of hooks on a line with the uppermost point of the humeral process; a dark area about dorsal fulcrum; a round dusky area at the base of each caudal lobe.

Anterior air-bladder with a bunch of finger-like cœca at the outer anterior angles; posterior air-bladder very small, forked, the two branches like a pair of horns, the points of the horns turned ventrad.

36. **Trachydoras paraguayensis** (Eigenmann and Ward).

Plate 1, Fig. 7; Plate 3, Fig. 8; Plate 18, Figs. 3 to 5; Plate 26, Fig. 1.

Hemidoras paraguayensis Eigenmann and Ward, Ann. Carnegie Mus., IV, 1907, p. 116, Pl. XXXIV, Fig. 1 (Corumba).

Hemidoras nattereri Fisher (in part), Ann. Carnegie Mus., XI, 1917, p. 421 (the specimens from San Joaquin, Corumba and Villa Hays).

Specimens Examined.

10127, I. U. M., *type*, 87 mm. to end of last scute. Anisits.

Part of 7201, C. M. 5, 46–88 mm. to end of middle caudal rays. San Joaquin, Bolivia. Haseman, Sept. 1909.

7203, C. M. one, 93 mm. over all. Villa Hays. Haseman, April 13, 1909.

Head 3.33–3.7; depth 3.25 to 3.33 (3 in Villa Hays specimen); D. I, 6; A. 14; scutes 3 + 29 or 30; eye 2 in snout, 4.5 in head, 2 in interorbital. Profile arched, occipital region steeply roof-shaped.

General appearance of *Doras brevis*. Head granular above, except for a space around and between the nostrils and the fontanel; fontanel not continued as a groove; opercle and preopercle granular; suborbitals very narrow, granular; preorbital granular; interopercle covered with skin; humeral process striate, a keel along its lower edge, the ridges near the upper edge stronger than those in the center; posterior edge emarginate, the upper and lower edges parallel; clavicle striate, united with the striate part of the coracoid into a hook-shaped area in front and below the pectoral spine; coracoid process striate, reaching as far as the humeral process; mouth narrow, inferior, maxillary barbels reaching to below some part of the eye, with 3 to 6 barblets; mental membrane narrow, the barbels papillose, without barblets. Teeth, if any, not evident; rudimentary scutes minute but each with a hook; first scute very broad, reaching from the dorsal plate to or near to the lower edge of the humeral process, the part above the hook in the adult with numerous thorns, in several irregular series; fewer in smaller; major hooks below the line of the hooks on the rudimentary scutes of the tympanum, about on a line with the upper margin of the humeral process; succeeding scutes much narrower and shorter, the series graduate to the last.

Dorsal spine shorter than pectoral spine, serrate in front to the upper third; the serræ on the hinder margin with a few minute points; adipose well developed, but usually shorter than the anal; anal more or less rounded; caudal very broad, the fin but slightly emarginate when expanded; pectoral spine reaching just beyond base of ventrals.

No markings; air-bladder heart-shaped; a small diverticulum branched or not at the outer anterior corners; the second air-bladder minute, forked, about equal to the diverticula.

37. **Trachydoras atripes** Eigenmann.

Plate 2, Fig. 5; Plate 18, Figs. 6 to 8; Plate 26, Fig. 4.

Specimens Examined.

15877, I. U. M., *type*, 60 mm. to end of lateral scutes, 72 mm. over all. Brook near R. Itaya, above Iquitos. Allen. Sept. 1920.

7204, C. M. 2, the larger about 80 mm. Berlin, Rio Marmoré. Haseman. Sept. 15, 1909.

7201, C. M., 16006, I. U. M. 9, 73 to 103 mm. to end of middle caudal rays. San Joaquin, Bolivia. Haseman. Sept. 1909.

Head 3.66 to 3.75; depth 4.25 to 5; D. I, 6; A. 14 to 17; scutes 3 + 34 or 35. Eye 1.5 to 2 in snout; 3 to 3.5 in head; 1 to 1.33 in interorbital; head granular, much as in *paraguayensis;* pit of the fontanel much longer, tapering at both ends, continued as a groove to the transverse groove on the occipital; profile not as steep as in *paraguayensis*, the occipital region roof-shaped; suborbital bones stronger and more prominently granular; nasal granular, its lower margin convex; interopercle and a small angle of lower jaw granular; mouth small, inferior, without evident teeth; maxillary barbel reaching to below the eye, with about 7 barblets; mental barbels united by a broad membrane, without barblets; rudimentary scutes variable, the last smallest, the two anterior linear or with the posterior half wider; coracoid processes and a linear streak of the coracoid, forward of the lower edge of the pectoral spine, granular; granular portion of the clavicle separated from the granular part of the coracoid, covered with skin; angle of the mandible exposed, granular; coracoid process considerably shorter than the humeral process; humeral process narrow, granular, rather than striate; first scute wider below the hook than above it, extending from the dorsal plate to a little below the humeral process, the surface with several sharp pointed granules, many marginal spines; scutes regularly graduate, line of hooks above on middle of the scutes, far above the upper margin of the humeral process; dorsal spine little if any shorter than the pectoral spine; serræ on posterior margin nearly as strong as on the anterior, the serræ antrorse but not as steeply so as on the anterior margin; adipose smaller than the anal; caudal deeply forked; anal rounded, its anterior half dusky.

Several branched cœca along the sides of the anterior air-bladder; posterior air-bladder very short, with branched cœca on each side.

In the type the dorsal fulcrum and base of dorsal rays are dark; anterior half of anal dark; pectoral dusky; middle caudal rays hyaline, a dusky band above and below the hyaline.

38. Trachydoras trachyparia (Boulenger).

Oxydoras trachyparia Boulenger, Trans. Zoöl. Soc. London, XIV, 1898, p. 423, Pl. XL, Fig. 2 (Rio Jurua).

This species is known by the two types, 92 mm. long, in the British Museum.

Genus XXI. Doras Lacépède.

Doras Lacépède, Hist. Nat. Poiss., V, 1803, p. 116; 12-mo., IX, 148 (*Silurus carinatus* Linnæus and *Silurus costatus* Linnæus).

Doras Bleeker, Nederl. Tijdsch. Dierk., I, 1863, p. 13 (type *carinatus*).

Oxydoras Kner, Sb. Ak. Wiss. Wien, XVII, 1855 (various species, the first *stenopeltis*, others *carinatus, niger* (*non* C. and V.), *lipophthalmus, d'Orbigny*, in the order named.

Hemidoras Bleeker, Ichthyol. Arch. Indici Siluri, 1858 (type *stenopeltis*). In 1862 Bleeker abandoned this generic name, placing *stenopeltis* in the present *Doras*.

Hemidoras Eigenmann and Eigenmann, Occasional Papers Cal. Acad. Sci., I, 1890, p. 250.

Oxydoras Günther, Cat. Fish. Brit. Mus., V, 1864, p. 206 (various species, the first one *carinatus*).

Doras Jordan, The Genera of Fishes, 1917, p. 65 (in listing the genera of Lacépède Jordan gives *Silurus carinatus* L. as the type).

Moryrostoma Ribeiro, Arch. Mus. Nac. Rio de Janeiro, 1911, p. 192 (*carinatus*).

Type, *Silurus carinatus* Linnæus.

This is not the *Doras* of Cuvier and Valenciennes, Günther, the Eigenmanns and authors, generally. Lacépède did not specifically state that he selected the *Silurus carinatus* of Linnæus as the type of his *Doras* but placed it first. Neither did Cuvier and Valenciennes specifically select a type for *Doras* but placed *costatus* first. Their restriction is not more specific than is that of Lacépède.

Bleeker specifically selected *carinatus* as the type of *Doras*. This seems to have been the first specific indication of the type. Jordan rules that *carinatus* being the first species mentioned by Lacépède is the type of his genus.

Mandibulary barbels with a basal membrane, with or without barblets; maxillary barbel with barblets; origin of the ventrals nearer the caudal than the snout; nuchal buckler intact, without foramen, without a median groove continued from the fontanel; the back between the dorsals without plates; lateral scutes moderate, the first one connecting the dorsal plate, the first rib and the humeral process; coracoid not granular; preopercle and interopercle covered with skin, not granular; a bony stay from epiotic to the lower end of the process of the dorsal plate, and first scute; adipose fin short, not continued as a keel.

Habitat: Rio Jauru to Guianas and Peru.

The air-bladders in this genus as accepted here differ considerably. In one type the air-bladder is simple without diverticula, excepting that in the old there is developed a minute second air-bladder, *carinatus, micropœus*. In a number of other species a second air-bladder is represented by two divergent recurved tubes. The species having this type are *brevis, punctatus, eigenmanni*, and possibly the adult of *microstomus*. In the third type it is heart-shaped with numerous radial diverticula. They are especially prominent in front.

KEY TO THE SPECIES OF DORAS.

a. Eye equal to or less than interorbital.

    b. First lateral scute unusually large, in contact with the coracoid process; teeth in both jaws; snout covered with skin; eye 1.5 in the snout, 1.25 in the interorbital; maxillary barbels reaching beyond base of pectoral; humeral process broad, rounded; a plate above and below the caudal peduncle; dorsal and pectoral spines of equal length, the latter extending beyond base of ventrals; scutes 29–30. Air-bladder simple, short, broad.......................................39. **fimbriatus** Kner.

    bb. First lateral scute not reaching coracoid process. Humeral process narrow.

        c. Dorsal and pectoral spines equal, the former reaching beyond origin of adipose; adipose fin equals anal; no teeth; origin of dorsal nearly midway between tip of snout and end of adipose; eye equals interorbital. Scutes 29–30; unspotted; depth 3.4; maxillary barbel reaching to below eye; mental barbels papillose; eye equals interorbital; air-bladder single, with very numerous diverticula and ending in two recurved horns...........................40. **brevis** Heckel.

        cc. Dorsal spine shorter than pectoral spine, not reaching the adipose.

d. Maxillary barbel reaching base of pectoral; mental barbels without slender barblets, papillose; snout considerably longer than the eye; scutes 29–30, the hook of the first one distinctly above the line of the rest; the line of hooks on a level with the upper margin of the humeral process; the rudimentary plates feeble, running below the middle of the pseudotympanum, the last one with a thorn, much above the first of the lateral scutes. Eye 1.66 in interorbital; center of opercle granular. Air-bladder ending in two recurved horns...41. **punctatus** Kner.

dd. Maxillary barbel reaching beyond base of pectoral; snout considerably longer than the eye; scutes 26 to 29; the hooks of the lateral scutes just above the line of the upper margin of the humeral process; eye 1.25 in interorbital. Mottled; mental barbels similar to the maxillary, with many barblets. Air-bladder ending in two recurved horns.

42. **eigenmanni** (Boulenger).

aa. Eye greater than interorbital.

e. Lateral scutes 28–32, rarely 26.

f. Maxillary barbel not reaching pectoral; snout less than the eye (in young only?); line of scutes distinctly above the level of the upper margin of the humeral process; the rudimentary plates rather strong, running at or a little above the level of the rest. Scutes 26–32. Interorbital less than vertical diameter of the eye, 1.5 in the longitudinal. Air-bladder in adult probably ending in two recurved horns..............................43. **microstomus** (Eigenmann).

ee. Lateral scutes 33–35 (Valenciennes gives 36); orbit elongate; dorsal spine shorter than pectoral spine; snout pointed; lateral plates narrow; interorbital much smaller than eye, 1.6–2 in the length of the eye. Young with a single air-bladder; old with a small second bladder.

g. Scutes well developed...44. **carinatus** (Linnæus).

gg. Anterior scutes minute; fins pigmented, dorsal dark, without a spot; snout long, subconical; the mouth very large; 4–10 teeth on each side of upper jaw, 2–5 on each side of lower jaw; no definite markings. Similar to *carinatus*...................45. **micropœus** (Eigenmann).

eee. Lateral scutes 37–38, largest on the base of the caudal; orbit much elongate; dorsal spine longer than pectoral spine; snout pointed, no teeth in upper jaw, lower jaw with minute teeth; barbels reaching gill-opening; the mental barbels with barblets similar to those of the maxillary; membrane of the mental barbels very wide; humeral process twice as long as high; interorbital equals vertical diameter of eye. Air-bladder heart-shaped, no second bladder or cœcum, numerous marginal diverticula, especially prominent in front.............................46. **lipophthalmus** Kner.

39. Doras fimbriatus Kner.

Doras fimbriatus Kner (*Corydoras loricatus* Heckel, MS.), Sb. Akad. Wiss. Wien, XVII, 1855, p. 134, Pl. III, Fig. 5, dorsal view of head (Rio Guaporé).

Oxydoras fimbriatus Günther, Cat. Fish. Brit. Mus., V, 1864, p. 207 (copied).

Hemidoras fimbriatus Eigenmann and Eigenmann, Occasional Papers Cal. Acad. Sci., I, 1890, p. 255.

Known from the types in the Vienna Museum.

40. Doras brevis Heckel.

Doras brevis [1] Heckel, MS. in Kner, Sb. Akad. Wiss. Wien, XVII, 1855, p. 138, Pl. VI, Fig. 11 (Barra do Rio Negro).

Known from the types in the Vienna Museum.

[1] The specimens listed as *brevis* by Fisher, Ann. Carnegie Mus., XI, 1917, p. 421, do not belong to this species.

41. Doras punctatus Kner.

Plate 3, Fig. 5; Plate 21, Figs. 1 to 2; Text-fig. 12, E.

Doras punctatus Kner (*Corydoras brevis* Heckel, MS.), Sb. Akad. Wiss. Wien, VII, 1855, p. 136, Pl. VI, Fig. 10
    (Matto Grosso; Rio Guaporé).

Corydoras punctatus Hyrtle, Denk. Ak. Wiss. Wien, XVI, 1859, p. 17 (vertebræ 5 + 1 + 21).

Oxydoras punctatus Günther, Cat. Fish. Brit. Mus., V, 1864, copied.

Hemidoras punctatus Eigenmann and Eigenmann, Occasional Papers Cal. Acad. Sci., I, 1890, p. 255.

Hemidoras nattereri Fisher, Ann. Carnegie Mus., XI, 1917, p. 421 (part of No. 7201, San Joaquin, Bolivia).

Hemidoras brevis Fisher, l.c., p. 421, part 7193 only (San Joaquin).

Specimens Examined.

7193, C. M., 96 mm. to end of middle caudal ray, and 7201, C. M. part 2, 69 to end of middle caudal ray.
    San Joaquin, Bolivia. Sept. 5, 1909. Haseman.

15882, I. U. M. one, 118 mm. R. Paranapura, Yurimaguas. Allen. Nov. 1920.

15883, I. U. M. many, largest 99 mm. Lake Cashiboya. Allen. Aug. 1920.

The specimens enumerated above differ from those described as *punctatus* by Kner in uniformly having longer maxillary barbels. Heckel had 17 specimens between 3 and 5 inches long from the Rio Guaporé. In these he says that the maxillary barbel uniformly fails to reach the gill-opening and has but 3 or 4 barblets.

Head 3.3 to 3.6; depth 3.9 to 4.2; D. I, 6; A. 12–14; scutes 2 + 1 [1] + 29; eye 1.7 in snout, 4 to 5 in the head, 1.6 to 1.75 in the interorbital. Head broad, narrowed to a sharp snout; occiput roofed; mouth subterminal, with teeth in two small patches in each jaw; opercle with a granular patch, cheeks and snout otherwise covered with skin; head above, behind posterior nostril finely granular-striate; fontanel elongate, ending abruptly above the middle of the eye, not continued as a groove; scapular region broad, the humeral process narrow, grooved and striate but not as deeply as in *eigenmanni;* its upper margin emarginate, its lower margin keeled; coracoid process as long as the humeral process; scutes in tympanum imbedded, *the last one with a hook* similar, but smaller than the hooks of the lateral series.

An inconspicuous spot near the center of the dorsal; an obscure dusky band along the middle of the back, from the dorsal to the caudal; a very obscure dusky band across the occipital and continued on the sides just above the hooks of the lateral line and above the center of the upper caudal lobe; a similar band through the center of the lower caudal lobe. In 15882, the bands are obsolete on the back and sides, but well marked on the caudal; the sides and the head in this specimen have rather sharp, small, dark spots. Posterior air-bladder forked.

[1] A rudimentary scute bearing a hook some distance above the hooks of the fully developed series.

42. Doras eigenmanni (Boulenger).

- Pate 21, Figs. 3 and 4.

Oxydoras eigenmanni Boulenger, Proc. Zoöl. Soc. London, 1895, p. 524; Trans. Zoöl. Soc. London, XIV,
1896, p. 28, Pl. IV, Fig. 3 (Descalvados, Matto Grosso); Eigenmann, McAtee and Ward, Ann.
Carnegie Mus., IV, 1907, 116 (Corumba).

Hemidoras brevis Fisher (part), Ann. Carnegie Mus., XI, 1917, p. 421 (Rio Jaurú).

Specimens Examined.

10140, I. U. M. 5, largest 107 mm. to end of middle caudal rays.   Corumba, Paraguay.   Anisits.

16190, I. U. M. 2, 78 and 85 mm.   Santarem market.   Carl Ternetz.

7190, C. M. 1, 95 mm. to end of middle caudal rays.   Santarem.   Dec. 15, 1909.   Haseman.

7191, C. M. 1, 80 mm. to end of middle caudal rays.   Caceres.   May 26, 1909.   Haseman.

7192, C. M. 1, 92 mm.   Maciél, Rio Guaporé.   Aug. 3, 1909.   Haseman.

7194, C. M. 3, 75–90 mm. to end of last scute.   Rio Jaurú.   June 4, 1909.   Haseman.

These are some of the specimens identified by Fisher as *brevis*.

Head 3.75 to 4; depth about equal to the head; D. I, 6; A. 12 to 13; scutes 28 to 29 (rarely 26).   Eye 2.5 in snout, 4.75 in head, 1.75 in interorbital.   Head broad, narrowed to a sharp snout; occiput roofed; mouth subterminal, with two small patches of comparatively strong teeth in both jaws; cheeks and snout in front of the posterior nostrils covered with skin, head above granular-striate from posterior nostril back; fontanel oval, ending abruptly over the pupil, not continued as a groove; scapular region broad, the humeral process deeply grooved and with prominent ridges, leaving a median, terminal depression free from keels; coracoid process heavy, about as long as the humeral process, finely striate from near the suture to the tip; scutes in the tympanum imbedded, scarcely evident; one scute only in contact with the dorsal plate; the series of hooks in a line just above upper margin of the humeral process, the hook of the first scute scarcely above the line of the rest; the scutes between a third and a fourth of the height, the median spine of the last not much smaller than those immediately in front of it; several marginal points above and below the median hook.

Maxillary barbel extending beyond the base of the pectoral spine; with about a dozen or more long, slender barblets in an alternating row on the ventral surface and outer edge; mental barbels and barblets (5–7) similar to the maxillary; basal membrane of mental barbels insignificant, the mental barbels reaching beyond the gill-opening, the outer considerably longer than the inner.   Dorsal spine strongly serrate on its anterior margin, less so on its posterior edge, its sides deeply grooved and ridged; pectoral spine usually extending beyond base of ventrals; origin of ventrals midway between snout and end of last scute, or nearer the later point; caudal forked;

anal rounded, adipose about as long as the anal. Everywhere mottled, light below; fins spotted.

This species resembles *punctatus* as figured by Heckel, in which, however, the first thorn of the lateral series lies higher than the rest and the maxillary barbels never reach the gill-opening and have but 3 to 4 barblets.

43. Doras microstomus (Eigenmann).

Hemidoras microstomus Eigenmann, Repts. Princeton Univ. Exp. Patagonia, III, 1910, p. 394 (name only); Mem. Carnegie Mus., V, 1912, p. 193, Pl. XVIII, Fig. 2 (Rockstone and Crab Falls, on the Essequibo, British Guiana); Steindachner, Flussf. Südam., V, 1915, p. 57 (Rio Branco at Boa Vista and Serra Grande; Rio Surumu).

Specimens Examined.

12039, I. U. M. Rockstone. Eigenmann.
12040, I. U. M. Crab Falls. Eigenmann.

Air-bladder broad, simple; an incipient double diverticulum at the end indicates that a forked double cœcum develops with age.

44. Doras carinatus (Linnæus).

Plate 1, Figs. 8, 9, 10; Plate 2, Fig. 6; Plate 20, Figs. 1 and 2; Plate 23, Fig. 4; Plate 27, Fig. 8.

Silurus carinatus Linnæus, Syst. Nat., ed. 12, I, 1766, p. 504; Bloch and Schneider, Syst. Ichth., 1801, p. 108.

Doras carinatus Lacépède, Hist. Nat. Poiss., V, 1803, p. 116 (Surinam); Cuvier and Valenciennes, Hist. Nat. Poiss., XV, 1840, p. 288, Pl. 442 (Cayenne); Müller and Troschel, in Schomburgk, Reisen, III, 1848, p. 629 (Essequibo); Bleeker, Ichth. Arch. Ind. Prodr., I, 1858, p. 54; Nederl. Tijdsch. Dierk., I, 1863, p. 13 (name only); Silures de Suriname, 1864, p. 31 (Surinam).

Doras (Oxydoras) carinatus Kner, Sb. Akad. Wiss. Wien, XVII, 1855, p. 144 (Surinam).

Oxydoras carinatus Günther, Catalogue, V, 1864, p. 206 (Surinam; Essequibo River); Vaillant, Bull. Soc. Philom. (7), IV, 1880, p. 154 (Calderon).

Hemidoras carinatus Eigenmann and Eigenmann, Proc. Cal. Acad. Sci. (2), I, 1888, p. 158; Occasional Papers Cal. Acad. Sci., I, 1890, p. 258; Eigenmann, Repts. Princeton Univ. Exp. Patagonia, III, 1910, p. 394; Mem. Carnegie Mus., V, 1912, p. 194.

Doras oxyrhynchus Valenciennes, in Humboldt, Rec. d'Obs. Zoöl. et Anat. Comp., II, 1833, p. 184 (Tumatumari on the Potaro; Bartica; Rockstone; Crab Falls; Georgetown market).

Specimens Examined.

12026, I. U. M. Rockstone. Eigenmann.
12027, I. U. M. Georgetown. Eigenmann.
12028, I. U. M. Crab Falls. Eigenmann.

45. Doras micropœus (Eigenmann).

Plate 1, Fig. 17; Plate 2, Fig. 1; Plate 20, Fig. 3; Plate 24, Figs. 1 and 2;
Plate 27, Fig. 5.

Hemidoras micropœus Eigenmann, Repts. Princeton Univ. Exp. Patagonia, III, 1910, p. 394 (name only);
  Mem. Carnegie Mus., V, 1912, p. 195 (British Guiana at Lama Stop-Off and Wismar on the Demerara
  River).
12029, I. U. M. Lama Stop-Off. Eigenmann.

46. Doras lipophthalmus Kner.

Doras (Oxydoras) lipophthalmus Kner (*Corydoras ophthalmus* Heckel, MS.), Sb. Akad. Wiss. Wien, XVII,
  1855, p. 147, Pl. 5, Fig. 8 (Rio Negro).
Oxydoras lipophthalmus Günther, Cat. Fish. Brit. Mus., V, 1864, p. 208 (River Capin).
Hemidoras lipophthalmus Eigenmann and Eigenmann, Proc. Cal. Acad. Sci., 2d Ser., I, 1888, p. 158; Occasional Papers Cal. Acad. Sci., I, 1890, p. 255.

This species is known from the types, 4 specimens 7½ inches long and a specimen
in the British Museum.

Genus XXII. HEMIDORAS Bleeker.

Hemidoras Bleeker, Ichthyol. Arch. Indici Siluri, 1858.

Type, *Hemidoras (Oxydoras) stenopeltis* Kner.

Bleeker who created the genus *Hemidoras* in 1858 abandoned it in 1862, placing
its type in the present genus *Doras*.

Snout long, subconical, smooth; lateral plates well developed along the entire
line; depth more than one fifth, rarely less than one fifth of the length; humeral
process longer than deep; mouth small; ventrals behind the middle of the body;
maxillary barbel with many barblets, mental barbels united by a basal membrane,
fringed with a double row of barblets.

A series of plates along the middle of the back between the dorsals and sometimes
between the anus and anal; a small foramen in the dorsal buckler; fontanel continued
as a groove.

The air-bladder in *morrisi* is prolonged in a short diverticulum to a point. Numerous branched tufts of diverticula around the entire margins.

KEY TO THE SPECIES OF HEMIDORAS.

a. Premaxillaries and dentaries without teeth or with a small patch of teeth; maxillary barbel extending
  beyond base of pectoral; scutes 32 to 37. Eye 1.5 in snout, 3–3.4 in head, greater than interorbital;
  D. I, 6...47. **stenopeltis** (Heckel).
aa. Eye (orbit) 4.25–5 in head, 2.25–2.75 in snout, 0.8 in interorbital; scutes 33–34.

48. **morrisi** Eigenmann.

47. Hemidoras stenopeltis (Heckel).

Corydoras stenopeltis Heckel, MS. as *Doras (Oxydoras) stenopeltis* in Kner, Sb. K. Akad. Wiss. Wien, XVII, 1855, p. 142, Pl. IV, Fig. 7 (Rio Negro).

Oxydoras stenopeltis Günther, Cat. Fish. Brit. Mus., V, 1864, p. 208 (copied).

Hemidoras stenopeltis Eigenmann and Eigenmann, Proc. Cal. Acad. Sci., 2d Ser., I, 1888, p. 158; Occasional Papers Cal. Acad. Sci., I, 1890, p. 256 (Manaos, Rio Negro; Hyavary; Manacapuru; Teffé; Obidos; Tabatinga); Fisher, Ann. Carnegie Mus., XI, 1917, p. 422 (Rio Madeira near San Antonio).

7186, C. M. 2, 107 and 132 mm. San Antonio. ' Nov. 3, 1909. Haseman.

15907, I. U. M. 1, 80 mm. Pebas, Peru. J. B. Steere.

Head 3.8 to 4; depth 5.2 to 5.33; D. I, 6; A. 13 to 15 (including rudimentary rays).

Eye 1.5 to 1.6 in snout, 3 to 3.4 in the head, about 0.75 in interorbital. Cheeks and snout without granulations; fontanel reaching the posterior edge of the pupil; skull striate; snout curved; mouth inferior; no teeth, or small patches of very minute teeth; anterior nostrils equidistant from tip of snout and eye; mental barbels with about 9 barblets, the basal half of which are scale-like, the posterior slender; maxillary barbel reaching beyond base of pectoral, with about 15 barblets along its outer edge, none on inner edge or surface. Three scutes on the tympanum, the first with downward projecting branch; 32 to 37 hook-bearing scutes, deepest in front, graduate to the last; numerous marginal spines both above and below the median hook; caudal peduncle with inconspicuous fulcra covering about half its length above and below; about ten plates along the back from the tip of the last dorsal ray, graduate to the adipose; dorsal spine as long as, or longer than, the pectoral spine, which reaches the ventrals; adipose fin nearly as long as the anal; caudal deeply forked; origin of ventrals little nearer snout than end of middle caudal rays; humeral process without distinct keel, its lower edge not parallel with the pectoral spine, the entire process below the line of the hooks of the lateral scutes.

A dark area about the dorsal fulcrum; tip of dorsal spine and of first ray black; a pair of dusky streaks on the caudal.

48. Hemidoras morrisi Eigenmann, spec. nov.

Plate 27, Fig. 3.

15962, I. U. M. 2, 77 and 147 mm., the latter the type. Iquitos. Morris.

Description of the Type: Head 3.2; depth about 6; D. I, 6; A. 14; scutes 33 or 34; distance between snout and end of dorsal plate $2\frac{3}{11}$ in the length; ventrals nearer tip of middle caudal rays than snout.

Slender, lateral scutes all well developed, about half the depth; the anterior ones with several (up to 6 or 7) marginal spines both above and below the median hook; about 14 very small scutes in front of the adipose; numerous, narrow, laminate caudal fulcra; a well-defined groove from the fontanel to the dorsal; snout smooth; suborbital line granular, upper posterior part of opercle granular, the rest of the opercle striate; two large granular scutes on the pseudotympanum; humeral spine slender, its dorsal margin emarginate; mouth without teeth (with up to 6 teeth on a side in each jaw in the paratype); snout long, slender; posterior nostril about equidistant from snout and eye; eye (orbit) slightly elongate, $\frac{2}{5}$ of the snout, $\frac{7}{5}$ in the interorbital, 4.33 in the head; maxillary barbel reaching about to base of pectoral; mental barbels short, their connecting membrane narrow; dorsal spine about equal to snout and eye (longer in paratype, equal to pectoral spine); adipose fin not quite as long as anal; pectoral spine almost equal to head, extending past origin of ventrals; a large pectoral pore. No prominent markings.

Genus XXIII.   OPSODORAS Eigenmann, gen. nov.

Type, *Opsodoras orthacanthus* Eigenmann, spec. nov.

Maxillary barbels with barblets, mental barbels with a double series of papils or barblets; a foramen on each side of the nuchal shield; no plates along the back; origin of the ventrals nearer caudal than snout; lateral scutes well developed, successive ones in contact, or the anterior ones minute, not in contact.

Air-bladder single, blunt or pointed behind, with fine, thread-like, branched diverticula, at least in *hemipeltis, parallelus, humeralis* and *orthacanthus*. In *leporhinus* of which only young were examined the air-bladder is without diverticula and ends in a blunt double point.

Habitat: Amazon and Guianas.

KEY TO THE SPECIES OF OPSODORAS.

a. Fontanel not continued as a groove to the dorsal plate; mental barbels papillose; well-developed teeth on the lower jaw; air-bladder profusely provided with thread-like cœca.

    b. Anterior scutes minute, successive ones not in contact with each other; interorbital convex; groove from fontanel reaching beyond eye; snout conical; its width in front of the eye little more than its depth; adipose eyelid prominent; maxillary barbel reaching to below posterior border of eye; a dark streak below the lower margin of eye.........................49. **hemipeltis** Eigenmann.

    bb. Depth of scutes about one half the diameter of the eye; foramen large, triangular; interorbital flat; groove of the fontanel truncate behind; snout depressed, much wider in front of eye than high; depth of humeral process at its widest less than the orbit; maxillary barbel reaching beyond the eye; scutes 0 + 32; edge and upper surface of the air-bladder most profusely supplied with cœca; pectoral spines not reaching ventrals.............................50. **parallelus** Eigenmann.

    bbb. Depth of scutes at least equal to the diameter of the eye, frequently much higher; foramen moderate, oval; groove of the fontanel pointed behind; depth of the humeral process at its widest equal to the orbit; scutes 3 + 31 or 32; air-bladders with numerous cœca; pectoral spines reaching ventrals.

51. **humeralis** (Kner).

aa. Fontanel continued to the dorsal plate as a well-defined, narrow groove.

 c. Keel of humeral process with its point continuous with the line of the hooks of the lateral scutes; scutes 30 to 31; eye 1 to 2 in interorbital; snout long, the distance between the nostrils about equal to the distance between the posterior and the eye; anterior nostril about equidistant from lip and the eye; mental barbels with a double row of barblets; foramen minute.

52. orthacanthus Eigenmann.

 cc. Keel and point of the humeral process below the line of hooks of the lateral scutes.

 d. A black spot on the base of the dorsal and another narrow oblique spot on the base of each caudal lobe.*

 e. The spots well defined; the lateral scutes about equal to the vertical diameter of the eye; maxillary barbel reaching the axil, with 3 short graduate barblets on the lower surface at inner edge of the barbel, near its base, with about 20 slender, graduate barblets along its outer edge; scutes 31–32. . . . . .53. **trimaculatus** (Boulenger).

 ee. Color markings more diffuse; the scutes with a light band, continued to end of middle caudal rays, dark above and below the band, the dark best marked on caudal; lateral scutes less than half the vertical diameter of the eye, 1.66 in its longitudinal; maxillary barbel reaching to pectoral spine, with about 11 flat, fringed barblets along its outer edge; mental barbels with numerous papils; each ramus of the lower jaw with about 5 brown-tipped teeth, those of the two rami forming a median group; lateral plates 33 to 35.

54. leporhinus (Eigenmann).

 dd. No distinct markings.

 f. Lateral scutes 29 to 35.

 g. Lower edge of humeral process convex; eye about equal to interorbital.

 h. Lateral scutes 35; eye a little greater than interorbital; mental barbels with barblets; the rudimentary scutes large; humeral process long, slender, three times as long as deep; depth of lateral scutes greater than orbit. . . . . . .55. **boulengeri** (Steindachner).

 hh. Lateral scutes 29 or 30, interorbital about equal to the eye; the rudimentary scutes minute; humeral process twice as long as deep; depth of scutes less than orbit.

56. stübeli (Steindachner).

 gg. Lower edge of humeral process straight; interorbital 1.3 in eye. Lateral scutes 32 to 35; dorsal spine equals pectoral spine, the latter reaching beyond the origin of the ventrals; eye about one fourth larger than the interorbital; scutes 34. . .57. **morei** (Steindachner).

 ff. Lateral scutes 38 to 40; eye three tenths larger than the interorbital.

58. steindachneri Eigenmann.

49. Opsodoras hemipeltis Eigenmann, spec. nov.

Plate 19, Fig. 2; Plate 24, Fig. 6.

15879, I. U. M., *type*, 143 mm. Rio Ucayali at Contamana. Allen. Sept. 1920.

Head 3.5; depth at occiput 6, at dorsal 5, at caudal peduncle 15.5; D. I, 6; A. 11; scutes about 33.

Sub-spindle shaped, the width everywhere approximating the depth; snout conical, opercles flush with the shoulders; interorbital convex; fontanel not continued

---

* A new species allied to *trimaculatus* and *leporhinus* has been received; see no. 53a.

as a groove, eye in center of the head, 1.5 in interorbital, 4 in the head; profile gently arched to the posterior nostrils, rapidly descending from the posterior nostrils forward; occipital striate, head otherwise smooth; snout small, width of the lower jaws 2 in the eye; barbels slender, basal membrane connecting the barbels small; the mental barbels simple, the maxillary barbels with a series of barblets along their outer edge, reaching to below posterior part of eye; width of isthmus equals snout; humeral process striate, triangular, its length equals snout, its height 1.5 in the eye; first eleven scutes rudimentary but increasing slightly, those above anal graduate, the most fully developed ones above end of anal; dorsal spine equals head behind posterior nostril, basal half of its anterior edge rough, its entire posterior edge serrate; adipose fin minute, half the length of the anal; caudal forked, its lobes equal; head behind posterior nostrils; anal short, truncate, its base less than snout; ventrals a little longer than snout, reaching half way to middle of anal; pectoral spines broken, the first ray equals head less half the snout; dusky above the lateral scutes, light below; lower border of orbit dark, no other markings.

Air-bladder prolonged, with many thin, branched diverticula.

50. Opsodoras parallelus Eigenmann, spec. nov.

Plate 19, Fig. 3; Plate 23, Fig. 3.

15964, I. U. M., *type*, 151 mm. Iquitos. Morris. 1922.

This species is evidently closely related, if distinct from *O. hemipeltis*. The lateral scutes are much better developed, the top of the head and nasal bones are granular, the opercle is striate; teeth of the lower jaw well developed but very fine, in two patches; mental barbels papillose, the maxillary barbels reaching to below middle of opercle; pectoral spines equal to the head behind the anterior nostrils, with small serræ on both surfaces to near the tip.

Profile less arched; interorbital flat.

Air-bladder prolonged into a point; with many thin, branched diverticula.

51. Opsodoras humeralis (Kner).

Plate 1, Fig. 6; Plate 2, Fig. 4; Plate 18, Figs. 9 and 10; Plate 19, Fig. 1.

Doras humeralis (*Corydoras humeralis* Heckel, MS.) Kner, Sb. Akad. Wiss. Wien, XVII, 1855, p. 140, Pl. IV, Fig. 6 (Barra do Rio Negro).
Oxydoras humeralis Günther, Cat. Fish. Brit. Mus., V, 1864, p. 206.
Hemidoras humeralis Eigenmann and Eigenmann, l.c., 1888, p. 158; 1890, p. 258.

Hitherto known only from the types.

Specimens Examined.

15880, I. U. M. one, about 154 mm. Rio Ucayali at Contamana. Allen. July 1920.

15881, I. U. M. 22, largest 145 mm. Specimen drawn 138 mm. long. Iquitos. Allen.

7189, C. M. 4, largest 122 mm. San Joaquin, Bolivia. Sept. 5 and 6, 1909. Haseman.

Resembling *brevis, leporhinus* and *microstomus.*

Head 3.5, depth at occiput 4.5–4.8; at dorsal spine 4–5; at caudal peduncle 14.5–15.5; D. I, 6; A. 10 to 12; scutes 0 to 4 + 29 to 32.

Heavy in front of dorsal; depth distinctly greater than the width except at caudal peduncle; snout subconical; opercles flush with the shoulders; interorbital slightly convex, fontanel not continued as a groove on the occipital; profile gently arched to posterior nostrils; the snout rapidly pointed in front of this point; distance between snout and anterior nostril about equal to the distance between the posterior nostril and eye, less than the distance between the anterior and posterior nostrils; top of head granular to posterior nostrils; opercle striate, head otherwise smooth; width of lower jaw a little more than half the length of the eyes; upper jaw with rarely a tooth; lower jaw with one or two series of teeth, those of the outer series larger, about 10 on each side; humeral process rough, very wide, its greatest width equal to the eye, 0.4–3 in its length, the margin from the point to its upper anterior angle either convex or with a distinct angle; mental barbels papillose, the maxillary barbel reaching to some point below the eye, with a lobe at its base on the inner margin, with about 7 slender barblets along its outer margin; scutes all well developed, graduated from the first back, about 0.4 in the height, with a median spine and a serrated margin; dorsal spine a little longer or shorter than the head; basal six tenths of anterior margin serrate, its entire posterior edge serrate; adipose fin half the length of the anal; caudal lobes about equal to the head behind the anterior nostrils; anal short, truncate, its base less than the snout; ventrals about equal to the snout; pectoral spines extending beyond origin of the ventrals; pectoral spine much longer than the dorsal spine, about equal to its distance from the snout to the upper angle of the humeral process; coracoid process covered.

Ashen above, light below; shadowy bands on the caudal lobes.

Air-bladder rounded behind, with many thin, branched marginal diverticula.

52. Opsodoras orthacanthus Eigenmann, spec. nov.

Plate 22, Fig. 3; Plate 23, Fig. 2.

Specimens Examined.

15884, I. U. M. 1, 133 mm. type. Iquitos. Morris. 1922.

15885, I. U. M. 4, 69–82 mm. Lake Cashiboya. Allen. Aug. 1920.

15886, I. U. M. 3, 77–98 mm. Mouth of Rio Pacaya. Allen. 1920.

Distinguished by the fact that the hooks on the lateral plates form a continuous, straight or nearly straight line with the ridge and tip of the humeral process.

Head 3.25; depth 5; D. I, 6; A. 12; scutes 30–31. Eye 2.33–2.66 in snout, 4.5–5 in the head, 1–2 in interorbital; snout acutely pointed; no teeth, width of lower jaw about 1.5 in the eye.

Width at the first dorsal and first anal rays equal to the depth at the same points; head wider than deep; head granular to the posterior nostrils; the foramen between dorsal plate and occipital minute; upper part of opercle granular; mental barbels with barblets; snout and cheeks covered with skin; humeral process triangular, the lower edge sloping, with a distinct ridge parallel with the depressed pectoral spine, its tip beyond the middle of the pectoral spine; coracoid process covered, reaching a little beyond middle of humeral process; dorsal spine serrate in front, roughened behind, about equal to snout and eye; pectoral spine equals length of head, reaching beyond origin of ventrals; base of adipose almost equal to base of anal; caudal forked, the upper lobe about equal to snout and eye; anal and ventrals rounded; two rudimentary scutes at beginning of lateral line at a much higher level than the fully developed ones; lateral scutes equally developed, the hooks under the dorsal much below the middle, behind the dorsal in the middle of the sides; each scute with a median hook and marginal serræ; sides above the hooks dusky, below them light, parallel dusky shades on the caudal lobes.

Air-bladder with many marginal diverticula.

53. Opsodoras trimaculatus (Boulenger).

Oxydoras trimaculatus Boulenger, Trans. Zoöl. Soc., XIV, 1898, p. 422, Pl. XL, Fig. 1 (Rio Jurua, 62 mm.).

Leptodoras trimaculatus Fowler, Proc. Acad. Nat. Sci. Phila., 1914, p. 264, Fig. 14 (Rupununi River, British Guiana, 77 mm.).

Hemidoras lipophthalmus (*non* Kner) Fisher, Ann. Carnegie Mus., XI, 1917, p. 421 (Santarem, 58 mm.).

Fowler proposed his *L. trimaculatus* as a new species but it is very probably identical with the *O. trimaculatus* of Boulenger.

7200, C. M. one, 58 mm. to end of middle caudal rays. Santarem. Haseman.

Head 3.66 in the length, caudal depth 4.6; D. I, 6; A. 14; scutes 2 or 3 + 30; eye 1 in snout, 2.8 in head, 0.5 in interorbital.

Snout very narrow, sharply decurved, pointed; mouth inferior, below the level of the eye, teeth very few, minute; cheeks and snout covered with skin, the skull striate, granular; a well-developed foramen on each side of the occiput; anterior nostril equidistant from snout and eye; fontanel very long, a bridge behind the pupil, continued as a groove to the dorsal plate; mental barbels very short, covered with wart-like papillæ; maxillary barbels reaching to some point on the base of the pectoral. four short graduate barblets on its inner lower surface, about 20 longer, more slender barblets along its outer edge; anterior scutes about 0.4 of the height, with about 4 marginal lobes below and one above the median hook; caudal fulcra numerous, laminate; lower edge of humeral process without a keel, its point below the line of the hooks; dorsal spine shorter than pectoral spine, longer than head; pectoral spine extending beyond the base of the ventrals; origin of ventrals a little nearer end of middle caudal rays than the snout; caudal deeply forked; a triangular black spot on base of anterior half of the dorsal, converging black stripes on base of caudal lobes.

This specimen is undoubtedly young. Its head is striate. It differs from the *trimaculatus* of Boulenger (62 mm.) in that the latter does not have the fontanel continued as a groove. It is possible that both are juvenile forms of a species resembling *O. steindachneri*.

53a. Opsodoras ternetzi Eigenmann, spec. nov.

16173, I. U. M. one specimen, type 128 mm. Tapajos at Santarem. 1924. Carl Ternetz.

Head 4; depth 4.75; D. I, 6; A. 10; plates 4 + 33.

Ventral profile very slightly curved; dorsal profile very convex to the dorsal, thence nearly straight but downward to the caudal peduncle. Snout conical; depth at dorsal fulcrum equal to length of snout and eye; eye 3.33 in head; eye behind middle of head; interocular 2 in length of eye; snout decurved; maxillary reaching axil, with about 17 barblets; fontanel continued to the dorsal plate; dorsal spine equal to length of head, with numerous spines, the antrorse spines extending the basal sixth of the spines, the retrorse spines extending about the basal six-sevenths of the posterior face of the spine; pectoral spine about equal to the dorsal spine, the hooks extending along the entire spine; caudal distinctly forked; depth of lateral scutes above the ventrals about 2 in length of eye. Upper surface of tip of snout dark; dorsal fulcrum dark; a narrow dusky streak on middle of back; a dark band above the scutes, a dusky streak below the scutes; a pair of dusky bands above and below the middle caudal rays. Lower margin and keel of humeral spine far below the line of the lateral spines; the small spines in the tympanum above the lateral hooks. Numerous and conspicuous mucous pores. Teeth not evident.

Air-bladder with irregular series of partly branched, lateral diverticula, a bunch of diverticula at middle of sides and another at the outer, anterior angle of the bladder; no posterior air-bladder, but an irregular protuberance from which diverticula radiate.

54. Opsodoras leporhinus (Eigenmann).

Hemidoras leporhinus Eigenmann, Repts. Princeton Univ. Exp. Patagonia, III, 1910, p. 394 (name only;)
Mem. Carnegie Mus., V, 1912, p. 195, Pl. XIX, Fig. 1 (British Guiana at Tumatumari, Potaro River,
and Crab Falls, Essequibo River); Steindachner, Flussf. Südam., V, 1815, p. 58, Pl. IX, Figs. 1–4
(Rio Surumu; Rio Branco at Serra Grande and Boa Vista).

I. U. M. 12021, 2 young, about 52 mm. Crab Falls. Eigenmann.

55. Opsodoras boulengeri (Steindachner).

Hemidoras (Leptodoras) boulengeri Steindachner, Flussf. Südam., V, 1915, p. 63, Pl. VIII, Figs. 1–3 (mouth
of Rio Negro).

Known from the types.

56. Opsodoras stübeli (Steindachner).
Plate 27, Figs. 6 and 7.

Oxydoras stübeli Steindachner, Flussf. Südam., IV, 1882, p. 5, Pl. III, Figs. 1–1b (Rio Huallaga).
Hemidoras stübeli Eigenmann and Eigenmann, Proc. Cal. Acad. Sci., 2d Ser., I, 1888, p. 158; Occasional
Papers Cal. Acad. Sci., I, 1890, p. 257.

Known from the types, three specimens 80–120 mm. long, in the Vienna Museum.

57. Opsodoras morei (Steindachner).
Plate 27, Figs. 1 and 2.

Oxydoras morei Steindachner, Flussf. Südam., II, 1881, p. 6, Pl. I, Figs. 2–2a (Rio Negro).
Hemidoras morei Eigenmann and Eigenmann, l.c., 1888, p. 158; 1890, p. 257.

Known from the type, a specimen 125 mm. long, in the Vienna Museum.

58. Opsodoras steindachneri Eigenmann, spec. nov.

Hemidoras carinatus non Linnæus, Steindachner, Flussf. Südam., V, 1915, p. 55 (mouth of Rio Negro).

Based on a specimen 197 mm. to base of caudal, 235 mm. over all, in the Vienna
Museum.

Head about $2\frac{2}{3}$; depth $4\frac{2}{7}$; scutes 38–40; D. I, 6; A. 3, 9.

Foramen tolerably large; snout long, compressed; maxillary barbels not quite
reaching pectoral; mental barbels connected for half their length by membrane;
mouth inferior, a few feeble teeth in each jaw; lateral plates low, of about uniform
height to the middle of the body, then increasing in height to the middle of the caudal
peduncle; depth about $4\frac{2}{7}$; distance between snout and dorsal 2.33 in the length;
humeral process oblique, with convex posterior margin, 3 times as long as high;
distance between tip of snout and ventrals $1\frac{7}{8}$ in the length.

Width of head 2 in its length; length of eye 4 in the length of the head, 2.35 in
snout; height of eye equals interorbital, 5.66 in length of head; dorsal spine $1\frac{1}{6}$,

pectoral 1⅖, ventral 1.8, depth of caudal peduncle 6, its length 2.4 in the length of the head.

The above is an abstract of Steindachner's description of *Doras carinatus* mentioned above. But *Doras carinatus* does not have a foramen except in the very young. It is therefore very probable that the specimen recorded by Steindachner, which has a large foramen, is not specifically or even generically identical with those recorded by Eigenmann from Guiana. Guiana specimens have about 33 scutes. Steindachner records 38 on one side and 40 on the other of his specimen. This makes Steindachner's identification doubly doubtful.

Genus XXIV. HASSAR Eigenmann and Eigenmann.

Hassar Eigenmann and Eigenmann, Proc. Cal. Acad. Sci., 2d Ser., I, 1888, p. 158.

Type, *Oxydoras orestis* Steindachner.

Snout, preopercle, and interopercle covered with skin. Anterior scutes of the lateral line very small; humeral process blunt, rounded, about twice as long as broad, extending past middle of pectoral spine, without keel; pectoral spine barely reaching ventrals or shorter; mouth large; a small patch of minute teeth on the lower jaw; orbit elongate; barbels slender, with barblets, with a narrow connecting membrane; ventrals behind the middle of the length; snout decurved, prolonged, pointed; first nostril equidistant from tip of snout and eye or nearer the latter; numerous small pectoral pores. A foramen on each side of x in the dorsal buckler.

Habitat: Amazon, Paranahyba, Guiana.

KEY TO THE SPECIES OF HASSAR.

a. Fontanel not continued as a narrow groove to the dorsal.

b. Dorsal plain; eye mostly in posterior part of head; adipose fin continued forward as a ridge, longer than anal...59. **affinis** (Steindachner).

bb. Dorsal with a black spot.

c. The black spot on the first three rays, not extending to their tips; eye extremely elongate; depth at base of pectoral equal to snout or snout and 0.4 of the eye; depth below dorsal spine 5.33–6 in the length to the last scute; head in the scapular region covered with skin, not very rough; distance between snout and dorsal about 2.5 in the length to the end of the lateral plates...60. **orestis** (Steindachner).

cc. The black spot on the first two rays, extending to their tip; [1] eye moderately elongate; depth at base of pectoral at least equal to snout and half the eye; depth below the pectoral spine about 4.66 in the length to the last scute; head in the scapular region more rugose; distance between snout and dorsal spine 2.5–2.6 in the length to the end of the lateral scutes.

61. **wilderi** Kindle.

aa. Fontanel continued as a narrow groove to the tip of the occipital process. A black spot on the dorsal; snout conical; dorsal spine with hooks on the basal part of the anterior margin; sides shaded toward the back; lateral plates 33..62. **notospilus** (Eigenmann).

[1] The dark area in Plate 29, Fig. 4 should be continued to the tip of the rays.

59. Hassar affinis (Steindachner).

Plate 27, Fig. 4.

Oxydoras affinis Steindachner, Flussf. Südam., II, 1881, p. 7, Pl. I, Fig. 1 (Rio Puty, tributary of Paranahyba north of Therezina).

Hemidoras affinis Eigenmann and Eigenmann, Proc. Cal. Acad. Sci., 2d Ser., I, 1888, p. 158; Occasional Papers Cal. Acad. Sci., I, 1890, p. 258 (Rio Puty); Steindachner, Flussf. Südam., V, 1915, p. 59 (Itapicurú at Caxias; Paranahyba at Engenho da Agua). .

60. Hassar orestis (Steindachner).

Oxydoras orestis Steindachner, Sb. Akad. Wiss. Wien, LXXI, 1875, p. 138, Pl. I (Xingu).

Hemidoras orestes Eigenmann and Eigenmann, Proc. Cal. Acad. Sci., 2d Ser., I, 1888, p. 158 (Huytahy); Occasional Papers, I, 1890, p. 258; Kindle, Ann. N. Y. Acad. Sci., VII, 1894, p. 251 (Itaituba, Brazil); Fisher, Ann. Carnegie Mus., XI, 1917, p. 422 (Santarem).

5123, I. U. M. three. Itaituba, Brazil. Hartt.

16172, I. U. M. nine, 143–255 mm. R. Tapajos, Santarem. C. Ternetz.

16188, I. U. M. four, 133–137 mm. Amazon, Santarem. C. Ternetz.

61. Hassar wilderi (Kindle).

Plate 22, Fig. 2.

Hassar wilderi Kindle, Ann. N. Y. Acad. Sci., VII, 1894, p. 251 (Troceras on Rio Tocantins).

Known from the types only, in the collections of Indiana University and of Cornell University.

5120 I. U. M., *types*, 162 and 207 mm. Troceras, Rio Tocantins.

62. Hassar notospilus (Eigenmann).

Hemidoras notospilus Eigenmann, Repts. Princeton Univ. Exp. Patagonia, III, 1910, p. 394 (name only); Mem. Carnegie Mus., V, 1912, p. 196, Pl. XIX, Fig. 2 (Crab Falls).

Known from the type, a specimen 70 mm. long, in the Carnegie Museum.

Genus XXV. LEPTODORAS Boulenger.

Leptodoras Boulenger, Ann. and Mag. Nat. Hist. (7), II, 1898, p. 478.

Type, *Oxydoras acipenserinus* Günther.

Air-bladder very short, its anterior end fitting into a cup-shaped downward process of the united vertebræ, two diverticula at posterior end, a pair of diverticula at the outer anterior corners.

Snout, preopercle and interopercle covered with skin. Slender, depth 5–9; humeral process not half as long as the pectoral spine, about as deep as long; mouth large, no teeth; barbels short, width of the membrane connecting the bases of the mental barbels more than half their length; maxillary barbels divided beyond the bone into an outer lateral fimbriated part and an inner part coterminous with the mental barbels; outer mental barbel divided distally into two; a small foramen in the dorsal buckler (or none?), a narrow groove extending back from the fontanel tending to disappear; lateral plates 36–44, the first scute extending down to the lower edge of the humeral process; in all but *hasemani* ventrals in front of the middle of the length; adipose sharply defined behind, continued as a low crest to near the tip of the depressed dorsal spine; snout decurved, the bones smooth; first nostril nearer the second than to the snout; one pectoral pore.

Habitat: Amazon, Guiana.

KEY TO THE SPECIES OF LEPTODORAS.

a. Dorsal plain.

   b. Dorsal spine with its membranous extension but little, if any, longer than the first ray; lateral plates covering half or more than half of the sides.

      c. Eye about 3.5–6 in the head; interorbital about half the length of the eye; distance between tip of snout and dorsal 3 or less than 3 in the length to the end of the lateral line; distance between the dorsals about equal to the length of the head; 38 or 39 lateral scutes; pectoral reaching middle of ventrals; depth 6–7; a pair of parallel, dusky bands on the caudal, the middle rays light . . . . 63. **linnelli** Eigenmann.

      cc. Eye 4.8–5 in the head, greater than interorbital; distance between tip of snout and dorsal more than 3 in the length to the end of the lateral line; distance between the dorsals much longer than the head; 42 lateral scutes; pectoral reaching base of ventral; depth 7.5.

                                                                                              64. **acipenserinus** (Günther).

   bb. Dorsal spine much produced beyond the soft rays, twice as long as the head; lateral plates covering less than half of the sides; eye 3 in snout, 1 in interorbital, 6.5 in head; maxillary barbel not reaching gill-opening; 44 lateral scutes; pectoral reaching base of ventral; depth 9.

                                                                                              65. **juruensis** Boulenger.

aa. Dorsal with a small round spot. Dorsal spine not longer than the following ray; lateral plates covering 0.4–0.5 of the body; eye a little more than 3.5 in the head; distance between snout and dorsal 2.75 in the length; distance between the dorsals equals length of head; 36–41 lateral scutes; pectoral not reaching middle of ventrals; depth 5–5.5; plain . . . . 66. **hasemani** (Steindachner).

63. Leptodoras linnelli Eigenmann.

Plate 2, Fig. 2; Plate 5, Figs. 3 and 4; Plate 20, Fig. 4; Plate 24, Figs. 3 to 5.

Leptodoras linnelli Eigenmann, Repts. Princeton Univ. Exp. Patagonia, III, 1910, p. 395 (name only); Mem. Carnegie Mus., V, 1912, p. 191, Pl. XVII, Fig. 1 and Pl. XVIII, Fig. 1 (British Guiana at Tumatumari on the Potaro; Rockstone and Crab Falls on the Essequibo River).

Leptodoras acipenserinus non Günther, Fisher, Ann. Carnegie Mus., XI, 1917, p. 422 (Maciél, Rio Guaporé).

Hemidoras (*Leptodoras*) *linnelli* Steindachner, Flussf. Südam., V, 1915, p. 65 (Rio Branco at Boa Vista and Serra Grande).

Specimens Examined.

7188, C. M. 3. Maciél, Rio Guaporé. Haseman.
12022, I. U. M. Tumatumari. Eigenmann.
12023, I. U. M. Crab Falls. Eigenmann.
12024, I. U. M. Rockstone. Eigenmann.

64. **Leptodoras acipenserinus** (Günther).

Oxydoras acipenserinus Günther, Proc. Zoöl. Soc. London, 1868, p. 230, Pl. XX (Xeberos); Steindachner, Flussf. Südam., II, 1881, p. 8 (Xeberos).

Hemidoras acipenserinus Eigenmann and Eigenmann, Proc. Cal. Acad. Sci., 2d Ser., I, 1888, p. 158; Occasional Papers Cal. Acad. Sci., I, 1890, p. 255.

Leptodoras acipenserinus Boulenger, Ann. and Mag. Nat. Hist. (7), II, 1898, p. 478.   (New genus.)

15878, one, 110 mm. Rio Paranapura at Yurimaguas. Allen. Nov. 1920.

Head 3.75; depth 8.5; width at origin of ventrals greater than depth at same place; depth of caudal peduncle 2 in its width with the spines; D. I, 6; A. 14; scutes 4 + 42.

Width of head equals the length of the snout, greater than its depth; eye 2.5 in snout, 4.8 in the head; interorbital 1.4 in the eye; lower profile straight; dorsal profile arched, descending from the nostrils to the sharp snout; fontanel continued as a narrow groove to the dorsal; top of head behind the eye granular; sides of head and snout with pearl-organ like papillæ; mouth inferior, narrow; width of lower jaw about 1.5 in the eye; snout projecting for nearly half the length of the eye; no teeth; bases of barbels connected by a broad membrane; maxillary barbel reaching below posterior part of eye; gill-openings wide, the isthmus equals length of eye; dorsal spine equals snout and half the eye, with antrorse spines on the basal half of its anterior edge, posterior margins roughened; adipose fin much shorter than the anal; caudal deeply forked, the lobes sharp pointed, a little greater than snout and eye; pectoral spines equal to snout and two thirds the eye; reaching past origin of ventrals, serrate on both margins, those on the posterior edge largest at the tip; scute less than half the depth, with a central spine and a serrated posterior edge; humeral process *half as broad as long*, reaching to middle of pectoral spines; coracoid process short, covered.   Back dusky; sides and lower parts lighter.   No distinct color markings.

Both this species and *linnelli* have a foramen where dorsal plate and occipital process meet.

65. **Leptodoras juruensis** Boulenger.

Leptodoras juruensis Boulenger, Ann. and Mag. Nat. Hist. (7), II, 1898, p. 478 (Jurua).

Known from the type only, a specimen 235 mm. long, in the British Museum.

66. Leptodoras hasemani (Steindachner).

Hemidoras hasemani Steindachner, Flussf. Südam., V, 1915, p. 61, X, Figs. 4–7 (Rio Negro at its mouth; Rio Branco at Boa Vista and Serra Grande).

Known from the types, the largest 132 mm. long, in the Vienna Museum.

Genus XXVI. NEMADORAS Eigenmann, gen. nov.

Type, *Oxydoras elongatus* Boulenger.

As far as known with the characters of *Opsodoras* but the maxillary barbels simple.　Fontanel not continued as a groove.

KEY TO THE SPECIES.

a. Depth 5 in the length; posterior nostril twice as distant from the anterior as from the eye; eye 4.5 in the head, 1.4 in interorbital; humeral process tapering to a point; pectoral spine as long as dorsal, nearly as long as head; D. I, 6; A. 12; 33 lateral scutes; olive above, white below; lateral shields and fins orange..67. **elongatus** (Boulenger).

aa. Depth 3.66 in the length; posterior nostril slightly nearer the eye than the anterior nostril; eye 2 in the snout, 4 in the head, 1.5 in interorbital; humeral process obliquely truncate posteriorly; pectoral spine longer than dorsal, a little longer than head; D. I, 5; A. 11; 30 lateral scutes; pale olive above, white below; fins white..68. **bachi** (Boulenger).

67. Nemadoras elongatus (Boulenger).

Plate XXI, Fig. 6.

Oxydoras elongatus Boulenger, Trans. Zoöl. Soc. London, XIV, 1898, p. 424, Pl. XL, Fig. 4 (Rio Jurua).

This species is known from the type, 105 mm. long, in the British Museum.

68. Nemadoras bachi (Boulenger).

Plate XXI, Fig. 5.

Oxydoras bachi Boulenger, l.c., p. 423, Pl. XL, Fig. 3 (Rio Jurua).

This species is known from the type, 90 mm. long, in the British Museum.

EXPLANATION OF PLATES.

General Notations on the Various Plates and Figures.—Notations peculiar to some one plate are given in the legend of the particular plate. For special notations for the auditory apparatus see legend of plate.

1 = Premaxillary; 2 = ethmoid; 3 = lateral ethmoid; 4 = preorbital (laminate); 5 = frontal; 6 = sphenotic; 7 = pterotic; 8 = supra-occipital; 8a = basi-occipital; 8b = ex-occipital; 9 = epiotic; 9a = process of the epiotic; 10 = supraclavicle; 11 = parapophysis of coalesced vertebræ; 12 = maxillary; 13 = palatine; 14 = mesopterygoid; 14a = metapterogoid; 15 = quadrate; 16 = preopercle; 17 = interopercle; 18 = opercle; 19 = hyomandibular; 20 = mandible; 21 = parasphenoid; 22 = prootic; 23 = orbitosphenoid; 23a = alisphenoid; *A. B. C.* = suborbitals; *D.* = nasal.

X, Y, Z = Elements of the dorsal plate.

A. IV = Anterior process of the fourth vertebra.

a = Anterior mesial articular process at base of dorsal spine.

b = Anterior articular surface of the first interneural.

c = Articular surface of the second interneural.

D. S. = Dorsal spine.

d = Opening through the base of the dorsal spine, through which the ring formed by the union of the pro-longations of the first and second interneural passes.

E = Modified first dorsal spine.

e = Opening admitting blood vessels, etc., into the interior of the dorsal spine.

f = Flanges or lateral processes of the interneurals.

K = Articular surface at the tip of the lateral flange of the second interneural.

l = Lateral articular surface at the base of the dorsal spine.

O = Modified first dorsal spine, see also *E*.

P. IV = Posterior process of the fourth vertebra.

Pr. IV = Process of the fourth vertebra.

tr = Tripus.

PLATE I.

Air-bladders of:

1. *Pterodoras granulosus* (Valenciennes), the ventral wall cut away exposing the inner surface of the dorsal wall.

2. The inner surface of the ventral wall, cut away from 1.

3. Dorsal view of the bladder of *Pterodoras granulosus*.

4. *Megalodoras irwini* Eigenmann. Dorsal view of the bladder of the adult.

5. *Platydoras costatus* (Linnæus), front view of air-bladder showing indentations into which the process of the fourth vertebra fits.

6. *Opsodoras humeralis* (Kner).

7. *Trachydoras paraguayensis* (Eigenmann and Ward).

8. *Doras carinatus* (Linnæus). Note the minute diverticulum at the middle of the lower end of the figure, the beginning of the second air-bladder.

9. *Doras carinatus* (Linnæus). The interior of the anterior end of the air-bladder, seen from immediately behind the cross-partition.

10. *Doras carinatus* (Linnæus). Longitudinal median section showing particularly the ridges and tentacles in the gullet and the air-bladder. The air-bladder shows the median partition "*m*," in the posterior part, the partial cross-partition "*c*," separating the posterior from the anterior part, the flexion over the fourth vertebra "*f*," and "*i*," the intruded end piece of the transverse process of the fourth vertebra.

11. *Astrodoras asterifrons* (Heckel).

12. *Centromochlus heckelii* Filippi, and

13. *Trachycorystes insignis* of the Auchenipteridæ.

14. *Platydoras costatus* (Linnæus).

15. *Oxydoras kneri* Bleeker.

16. *Pseudodoras niger* (Valenciennes).

17. *Doras micropœus* (Eigenmann).

18. *Rhinodoras d'orbignyi* (Kröyer).

PLATE II.

FIG. 1. *Doras micropœus* (Eigenmann). I. U. M., 12029, 245 mm., Lama Stop-Off.

FIG. 2. *Leptodoras linnelli* Eigenmann. I. U. M., 12022, about 180 mm., Tumatumari.

FIG. 3. *Acanthodoras spinosissimus* (Eigenmann and Eigenmann). I. U. M., 16005, 63 mm., Maciél.

FIG. 4. *Opsodoras humeralis* (Kner). I. U. M., 15881, about 120 mm., Iquitos.

FIG. 5. *Trachydoras atripes* Eigenmann. I. U. M., 16006, about 100 mm., S. Joaquin.

FIG. 6. *Doras carinatus* (Linnæus). I. U. M., 12026, about 185 mm., Rockstone.

FIG. 7. *Lithodoras dorsalis* (Valenciennes). I. U. M., 4248, about 165 mm., Pará.

FIG. 8. *Centromochlus heckelii* Filippi. Auchenipteridæ.

FIG. 9. *Trachycorystes insignis* (Steindachner). Auchenipteridæ.

PLATE III.

FIG. 1. *Platydoras costatus* (Linnæus). I. U. M., 77 mm., Rio Morona.

FIG. 2. *Platydoras costatus* (Linnæus). I. U. M., 17041, 185 mm., Lake Rogoagua.

FIG. 3. *Megalodoras irwini* Eigenmann. I. U. M., 15427, about 300 mm., now shrunken, Iquitos.

FIG. 4. *Centrochir crocodili* (Humboldt). I. U. M., 13557, about 175 mm., Soplaviento.

FIG. 5. *Doras punctatus* Kner. I. U. M., 15883, about 95 mm., Lake Cashiboya.

FIG. 6. *Pterodoras granulosus* (Valenciennes). I. U. M., 15847, 165 mm. to end of middle caudal ray, Iquitos.

FIG. 7. *Franciscodoras marmoratus* (Reinhardt). After Steindachner.

FIG. 8. *Trachydoras paraguayensis* (Eigenmann and Ward). I. U. M., 16009, 81 mm. to end of middle caudal rays, S. Joaquin.

FIG. 9. *Amblydoras hancocki* (Cuvier and Valenciennes). I. U. M., 12034, about 95 mm., Maduni Creek.

PLATE IV.

FIG. 1. *Hypodoras forficulatus* Eigenmann. I. U. M., 15876, *type*, 123 mm., Iquitos.

FIG. 2. *Astrodoras asterifrons* (Heckel). I. U. M., 16003, 97 mm., Jutahy.

FIG. 3. *Scorpiodoras heckeli* (Kner). I. U. M., 16002, 161 mm., Teffé.

FIG. 4. *Astrodoras asterifrons* (Heckel). I. U. M., 16035, 80 mm., Santarem.

FIG. 5. *Acanthodoras cataphractus* (Linnæus). I. U. M., 12032, about 82 mm., Gluck Island.

FIG. 6. *Anadoras weddelii* (Castelnau). I. U. M., 5110, 113 mm., Marajo.

FIG. 7. *Anadoras grypus* (Cope). I. U. M., 15849, about 120 mm., Lake Cashiboya.

PLATE V.

Radiographs by Dr. J. E. P. Holland.

FIGS. 1 AND 2. *Platydoras costatus* (Linnæus).
FIGS. 3 AND 4. *Leptodoras linnelli* Eigenmann.
FIGS. 5 AND 6. *Centrochir crocodili* (Humboldt).

PLATE VI.

Photographs of *Megalodoras irwini* Eigenmann. I. U. M., 15427, about 300 mm.
FIG. 1. Membrane bones of left side from without.
FIG. 2. Dorsal view.
FIG. 3. Suborbitals and preorbital.
FIG. 4. Ventral view of the skull and membrane bones.

PLATE VII.

Megalodoras libertatis (Ribeiro).

Drawings of the type by E. Cruzlima. Loaned by Dr. A. Ribeiro.

PLATE VIII.

Photographs of *Pterodoras granulosus* (Valenciennes).

FIG. 1. Side view. I. U. M., 15848, 180 mm., R. Puinahua.
FIG. 2. Dorsal view of the same.
FIG. 3. Side view of another specimen, skeletonized.
FIG. 4. Ventral view of skull and membrane bones.

PLATE IX.

FIG. 1. Side view of a young specimen of *Platydoras costatus* (Linnæus). I. U. M., 15874, 82 mm., Lake Yarinococha.
FIG. 2. Dorsal view of an older specimen.
FIG. 3. Side view of skeleton. I. U. M., 17041, Lake Rogoagua.
FIG. 4. Ventral view of *Centrochir crocodili* (Humboldt). 13557, about 175 mm., Soplaviento. See Plate III, Fig. 4.
FIG. 5. *Centrochir crocodili* (Humboldt). From Steindachner.
FIG. 6. *Lithodoras dorsalis* (Valenciennes). I. U. M., 4248 mm., about 165 mm., Pará. See Plate II, Fig. 7.

PLATE X.

FIGS. 1 AND 2. *Acanthodoras cataphractus* (Linnæus). I. U. M., 12032, 86 mm., Gluck Island.
FIGS. 3 AND 4. The same from a specimen 80 mm.
FIG. 5. *Acanthodoras calderonensis* (Vaillant). After Steindachner.

PLATE XI.

FIGS. 1 AND 2. *Acanthodoras spinosissimus* Eigenmann and Eigenmann. C. M., 7167, Maciél.
FIGS. 3 AND 4. The same. I. U. M., 16005, 93 mm., Maciél.

PLATE XII.

FIG. 1. *Megalodoras uranoscopus* Eigenmann and Eigenmann. M. C. Z. type, 530 mm. E. N. Fischer, del.

FIG. 2. Dorsal view of the same specimen.

PLATE XIII.

Amblydoras hancockii (Cuvier and Valenciennes).

FIGS. 1 AND 2. I. U. M., 12034, 118 mm., Maduni Creek.

FIG. 3. I. U. M., 12033, 112 mm., Lama Stop-Off.

FIG. 4. I. U. M., 12034, 95 mm., Maduni Creek.

PLATE XIV.

Anadoras insculptus (Ribeiro).

Drawings by E. Cruzlima, based on the type in the Museu Nacional, Rio de Janeiro. Dr. A. Ribeiro has kindly loaned these drawings.

PLATE XV.

FIGS. 1–3. *Anadoras grypus* (Cope). 1 and 2, I. U. M., 15849, 139 mm., Lake Cashiboya. 3, 118 mm., Lake Cashiboya.

FIG. 4. *Anadoras weddelii* (Castelnau). I. U. M., 5110, 113 mm., Marajo.

PLATE XVI.

FIG. 1. Ventral view, 2. dorsal view of *Physopyxis lyra* Cope. *Type* in Mus. Phila. Acad. Sci.

FIG. 3. *Amblydoras monitor* (Cope). *Type* in Mus. Phila. Acad. Sci.

FIG. 4. *Agamyxis pectinifrons* (Cope). *Type* in Mus. Phila. Acad. Sci.

PLATE XVII.

FIGS. 1–4. *Pseudodoras niger* (Valenciennes). Figs. 1 and 2, side and dorsal views of a specimen I. U. M. 16004, 273 mm. to end of the scutes, Rio Pacaya. Allen. Figs. 3 and 4, I. U. M., 15651, about 237 mm., Iquitos.

PLATE XVIII.

FIGS. 1 AND 2. *Trachydoras nattereri* (Steindachner). I. U. M., 15967, 131 mm., Iquitos.

FIGS. 3–5. *Trachydoras paraguayensis* (Eigenmann and Ward). Fig. 3, C. M., part of 7201, San Joaquin. Figs. 4 and 5, I. U. M., 16009, 81 mm. to end of middle caudal rays, San Joaquin.

FIGS. 6–8. *Trachydoras atripes* Eigenmann. Figs. 6 and 7, I. U. M., 16006, about 100 mm., San Joaquin. Fig. 8, C. M., part of 1201, San Joaquin.

FIGS. 9–10. *Opsodoras humeralis* (Kner). I. U. M., 15881, about 120 mm., Iquitos.

PLATE XIX.

FIG. 1. *Opsodoras humeralis* (Kner). I. U. M., 15881, 138 mm., Iquitos. W. S. Atkinson, del.

FIG. 2. *Opsodoras hemipeltis* Eigenmann. I. U. M., 50879, *type*, 143 mm., Contamana. W. S. Atkinson, del.

FIG. 3. *Opsodoras parallelus* Eigenmann. I. U. M., 15964, *type*, 151 mm., Iquitos. W. S. Atkinson, del.

PLATE XX.

FIGS. 1–2. *Doras carinatus* (Linnæus). I. U. M., 12026, about 185 mm., Rockstone.

FIG. 3. *Doras micropœus* (Eigenmann). I. U. M., 12029, 245 mm., Lama Stop-Off.

FIG. 4. *Leptodoras linnelli* Eigenmann. I. U. M., 12022, about 180 mm., Tumatumari.

PLATE XXI.

FIGS. 1 AND 2. *Doras punctatus* Kner. I. U. M., 15882, 118 mm., R. Paranapura, Yurimaguas.

FIGS. 3 AND 4. *Doras eigenmanni* (Boulenger). After Boulenger.

FIG. 5. *Nemadoras bachi* (Boulenger). After Boulenger.

FIG. 6. *Nemadoras elongatus* (Boulenger). After Boulenger.

PLATE XXII.

FIG. 1. *Franciscodoras marmoratus* (Reinhardt). After Lütken.

FIG. 2. *Hassar wilderi* Kindle. I. U. M., 5120, *type*, about 165 mm., Troceras. W. S. Atkinson, del.

FIG. 3. *Opsodoras orthacanthus* Eigenmann. I. U. M., 15884, *type*, 133 mm., Iquitos. W. S. Atkinson, del.

PLATE XXIII.

FIG. 1. *Pseudodoras niger* (Valenciennes). T. R. S., No. 743, 840 mm., Koriabo, Guiana. Photo by John Tee-Van of a giant specimen from Kartabo.

FIG. 2. *Opsodoras orthacanthus* Eigenmann. I. U. M., 15884, *type*, 133 mm., Iquitos. See Plate 22, Fig. 3.

FIG. 3. *Opsodoras parallelus* Eigenmann. I. U. M., 15964, *type*, 151 mm., Iquitos. See Plate 19, Fig. 3.

FIG. 4. *Doras carinatus* (Linnæus). I. U. M., 12026, 233 mm. to end of middle caudal rays, Rockstone.

PLATE XXIV.

FIGS. 1 AND 2. *Doras micropœus* (Eigenmann). Fig. 2, I. U. M., 12029, 245 mm., Lama Stop-Off.

FIGS. 3–5. *Leptodoras linnelli* Eigenmann. Fig. 3, I. U. M., 12022, 167 mm., Tumatumari. Fig. 4, The same specimen from below. Fig. 5, I. U. M., 12022, about 180 mm., Tumatumari. See Plate 2, Fig. 2.

FIG. 6. *Opsodoras hemipeltis* Eigenmann. I. U. M., 15964, *type*, 151 mm., Iquitos.

PLATE XXV.

FIG. 1. *Megalodoras irwini* Eigenmann. T. R. S., 2567, 612 mm., Kartabo.

FIG. 2. *Megalodoras irwini* Eigenmann. I. U. M., 15427, *type*, 278 mm., Iquitos. W. S. Atkinson, del.

FIG. 3. *Hypodoras forficulatus* Eigenmann. I. U. M., 15876, *type*, 123 mm., Iquitos. W. S. Atkinson, del.

PLATE XXVI.

FIG. 1. *Trachydoras paraguayensis* (Eigenmann and Ward). I. U. M., 10127, *type*, 87 mm., Corumba. W. S. Atkinson, del.

FIGS. 2 AND 3. *Trachydoras nattereri* (Steindachner). After Steindachner.

FIG. 4. *Trachydoras atripes* Eigenmann. I. U. M., 15877, *type*, 60 mm. to end of lateral scutes, Iquitos. W. S. Atkinson, del.

FIG. 5. *Rhinodoras d'orbignyi* (Kröyer). I. U. M., 9837, *type* of *Doras nebulosus* Eigenmann and Ward, 160 mm., Asuncion. W. S. Atkinson, del.

PLATE XXVII.

FIGS. 1 AND 2. *Opsodoras morei* (Steindachner). After Steindachner.

FIG. 3. *Hemidoras morrisi* Eigenmann. I. U. M., 15962, *type*, 147 mm., Iquitos. W. S. Atkinson, del.

FIG. 4. *Hassar affinis* (Steindachner). After Steindachner.

FIG. 5. *Doras micropæus* (Eigenmann). I. U. M., 12029, about 355 mm., Lama Stop-Off. W. S. Atkinson, del.

FIGS. 6–7. *Opsodoras stübeli* (Steindachner). After Steindachner.

FIG. 8. *Doras carinatus* (Linnæus). I. U. M., 12025, about 235 mm., Tumatumari. W. S. Atkinson, del.

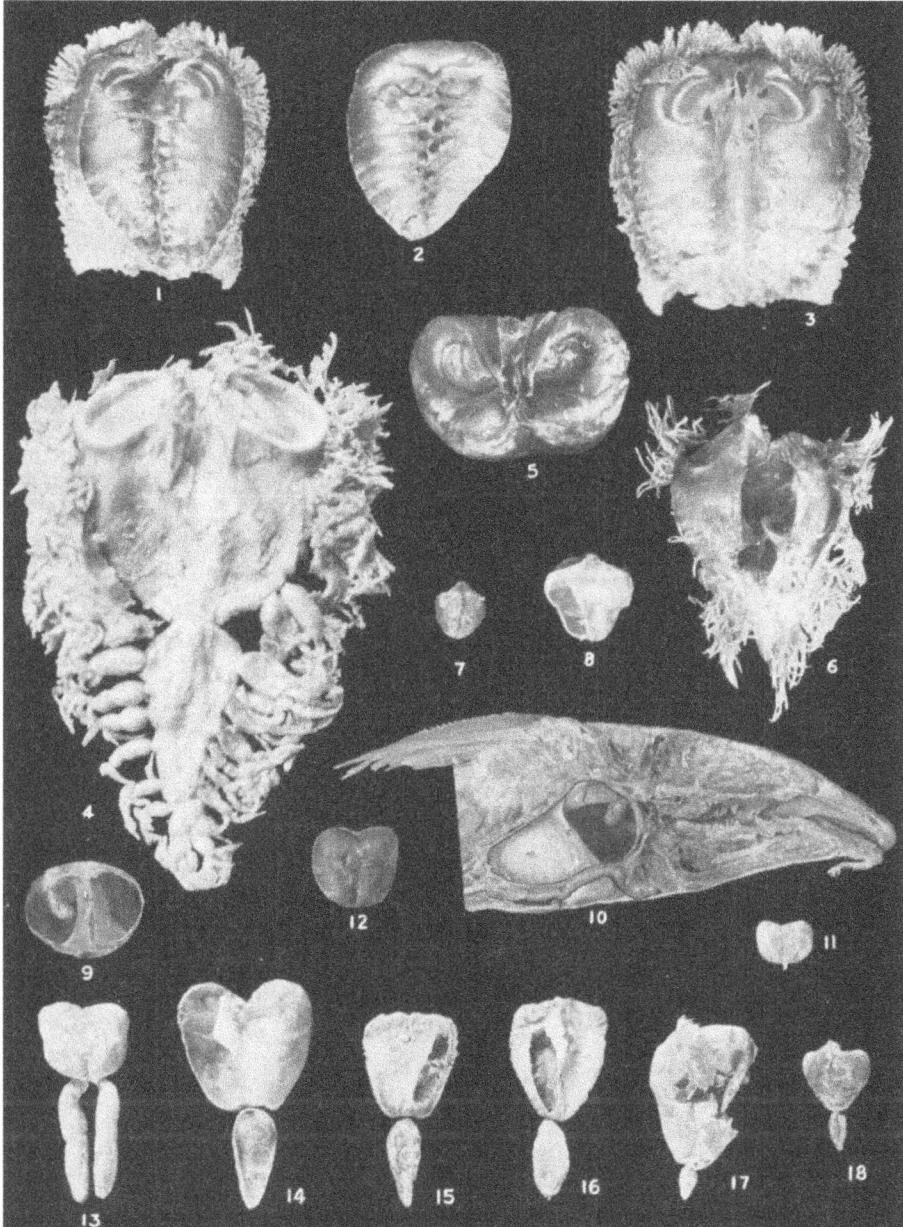

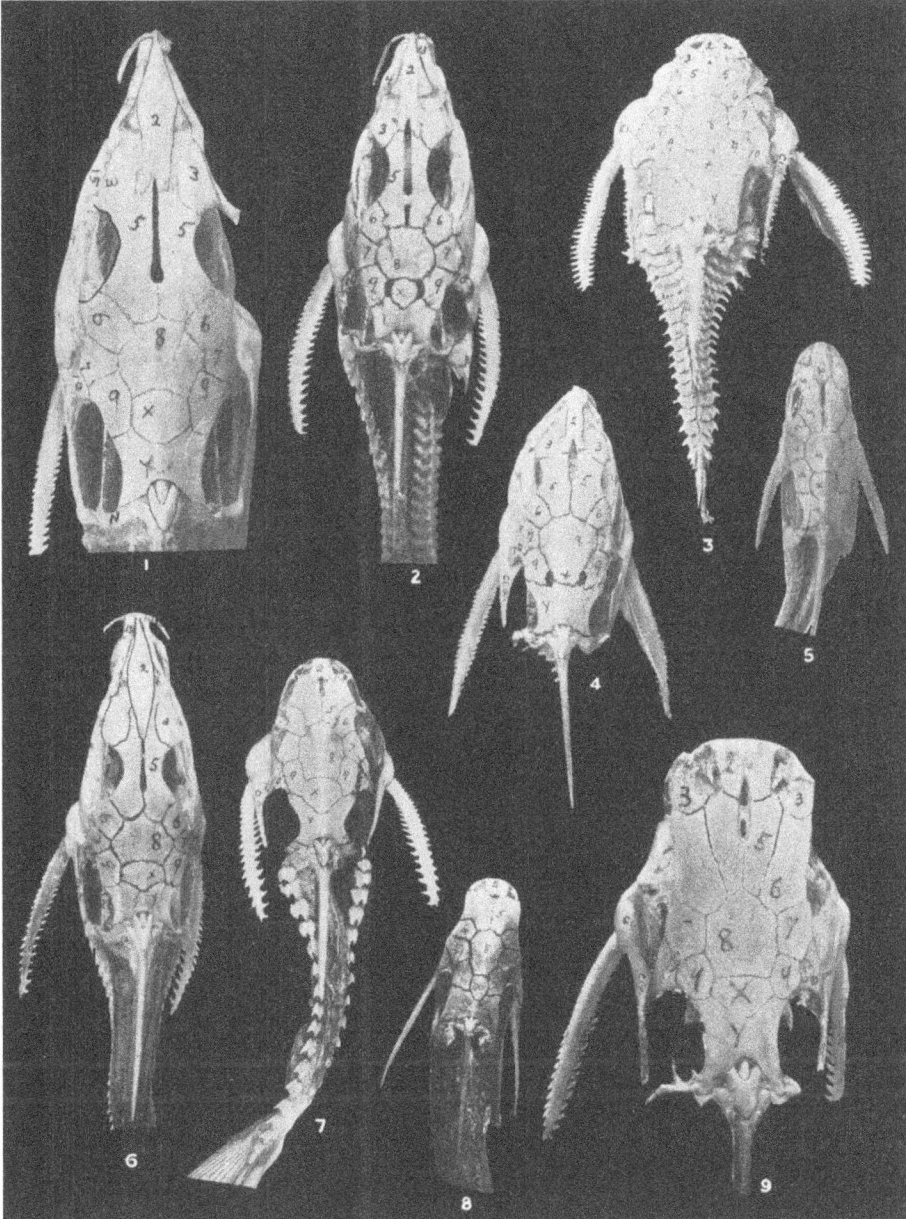

PLATE III.

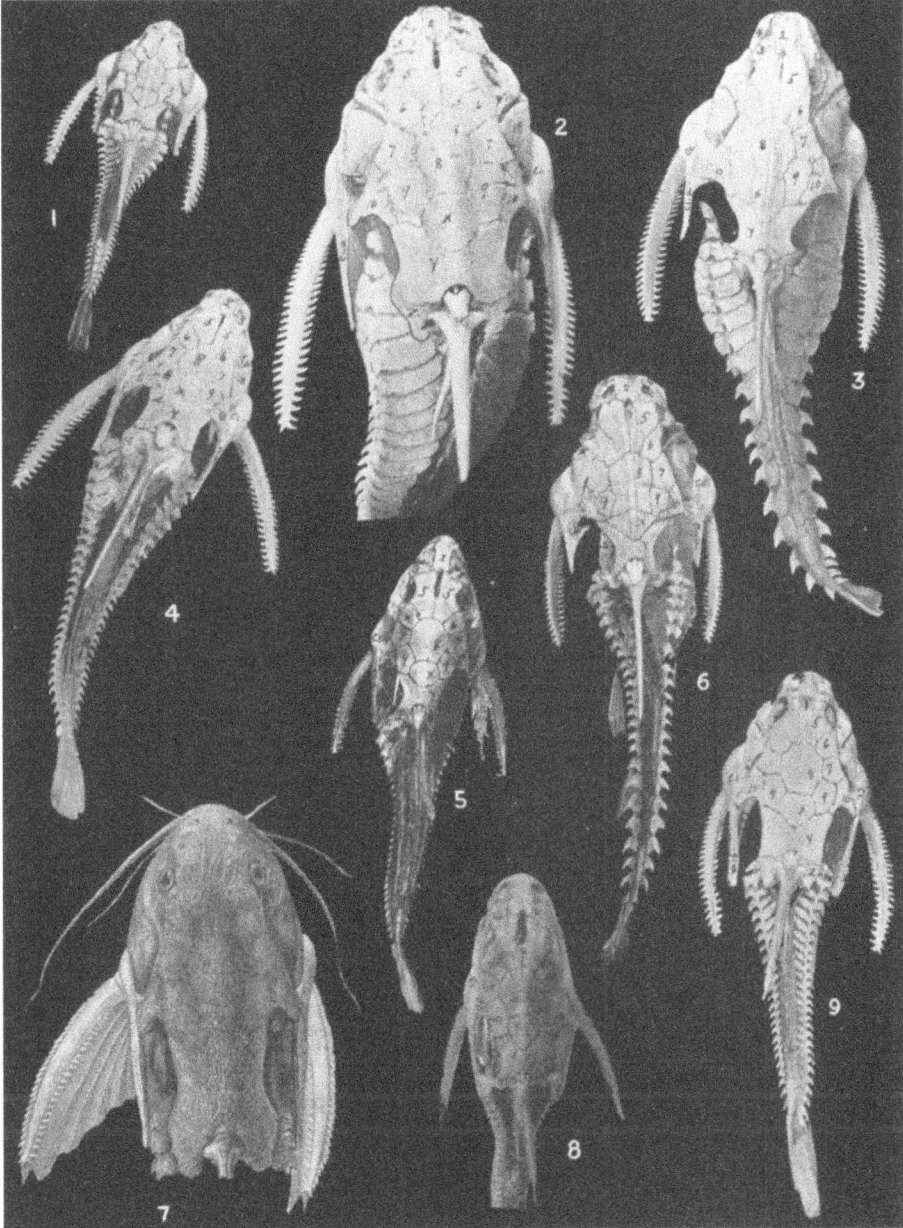

PLATE IV.

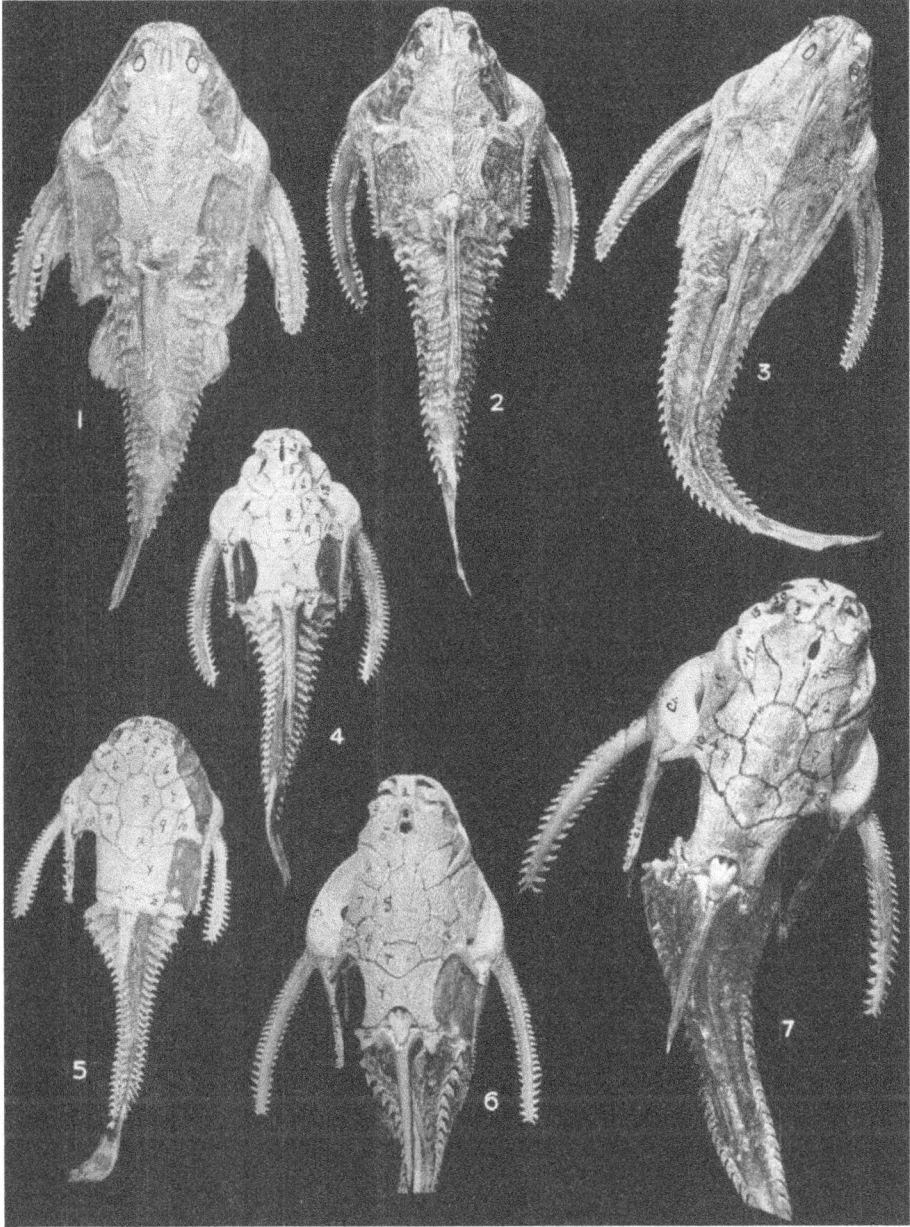

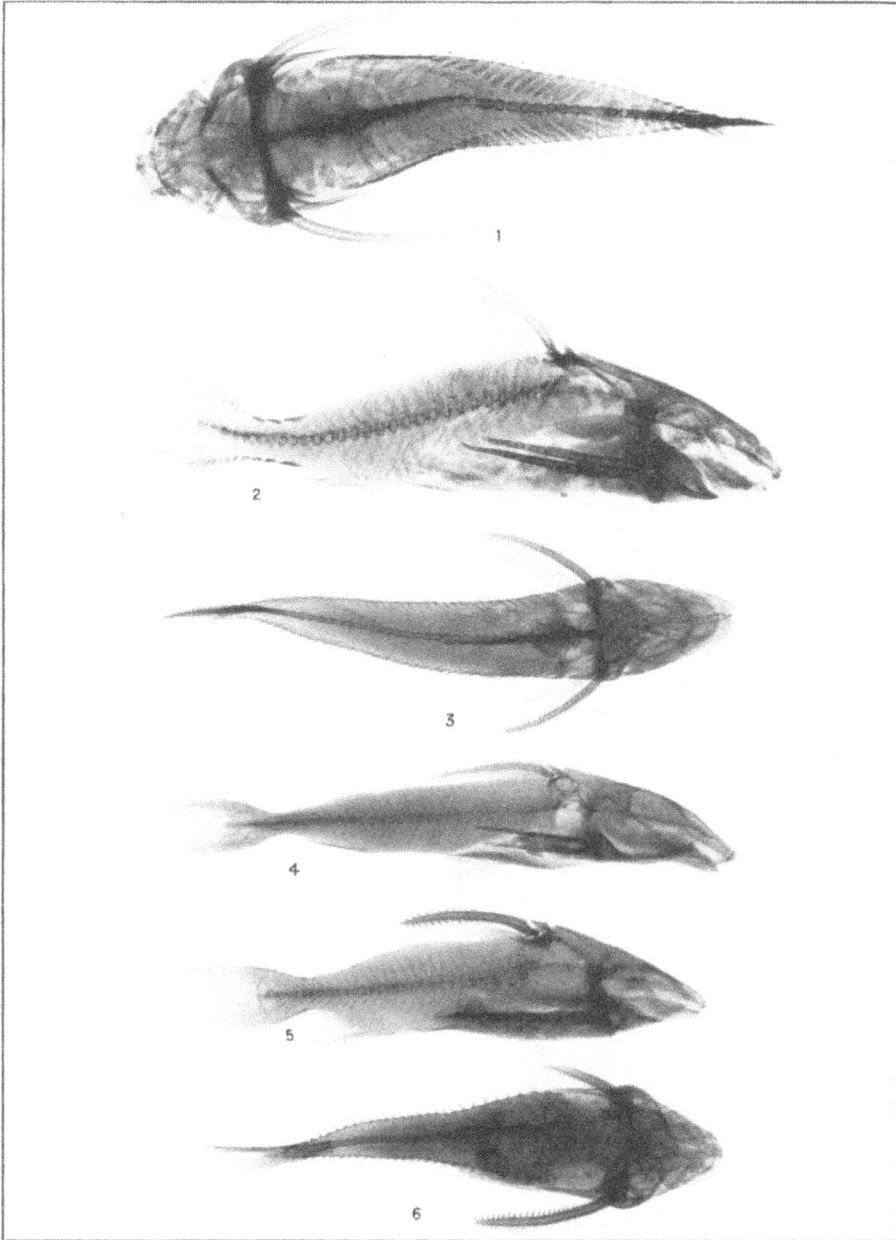

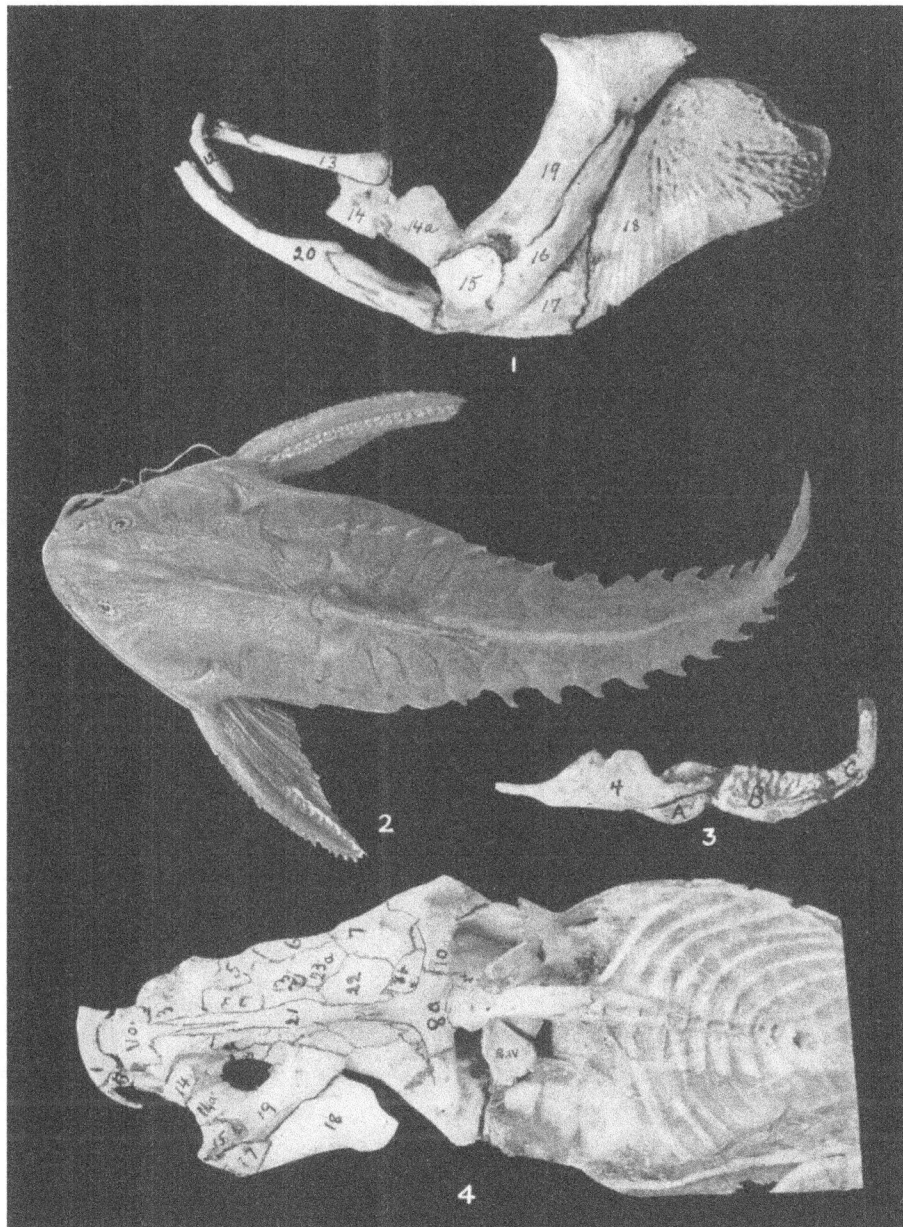

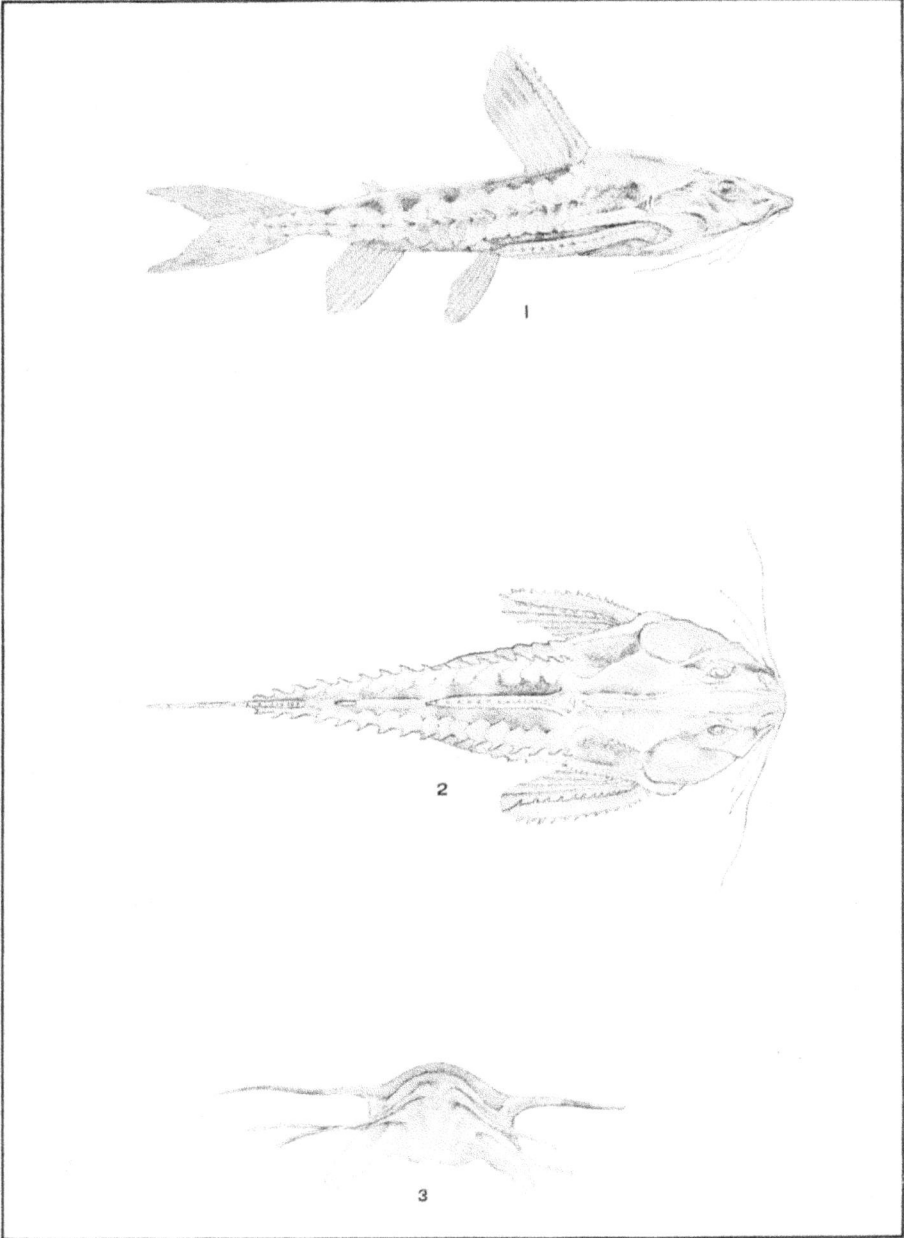

1

2

3

PLATE VIII.

PLATE IX.

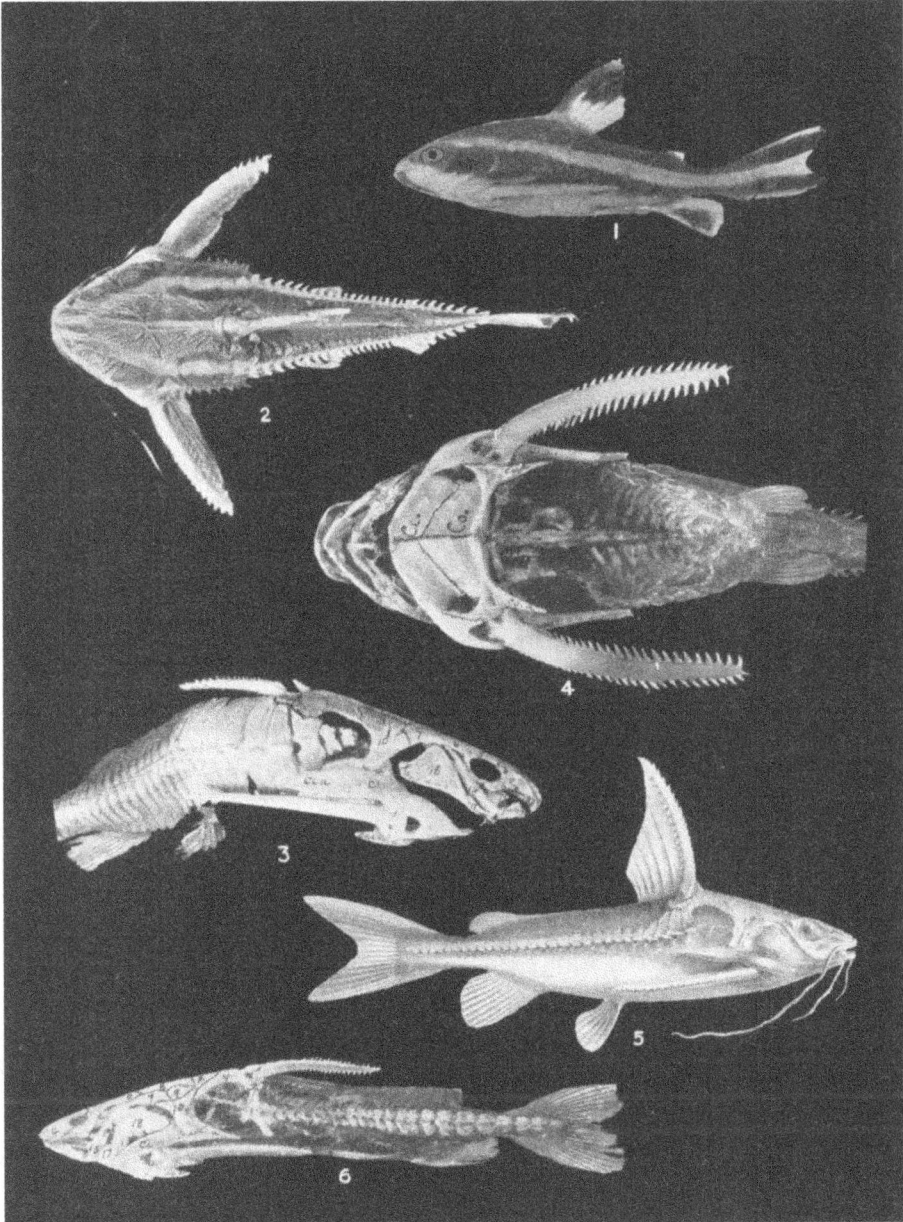

PLATE X.

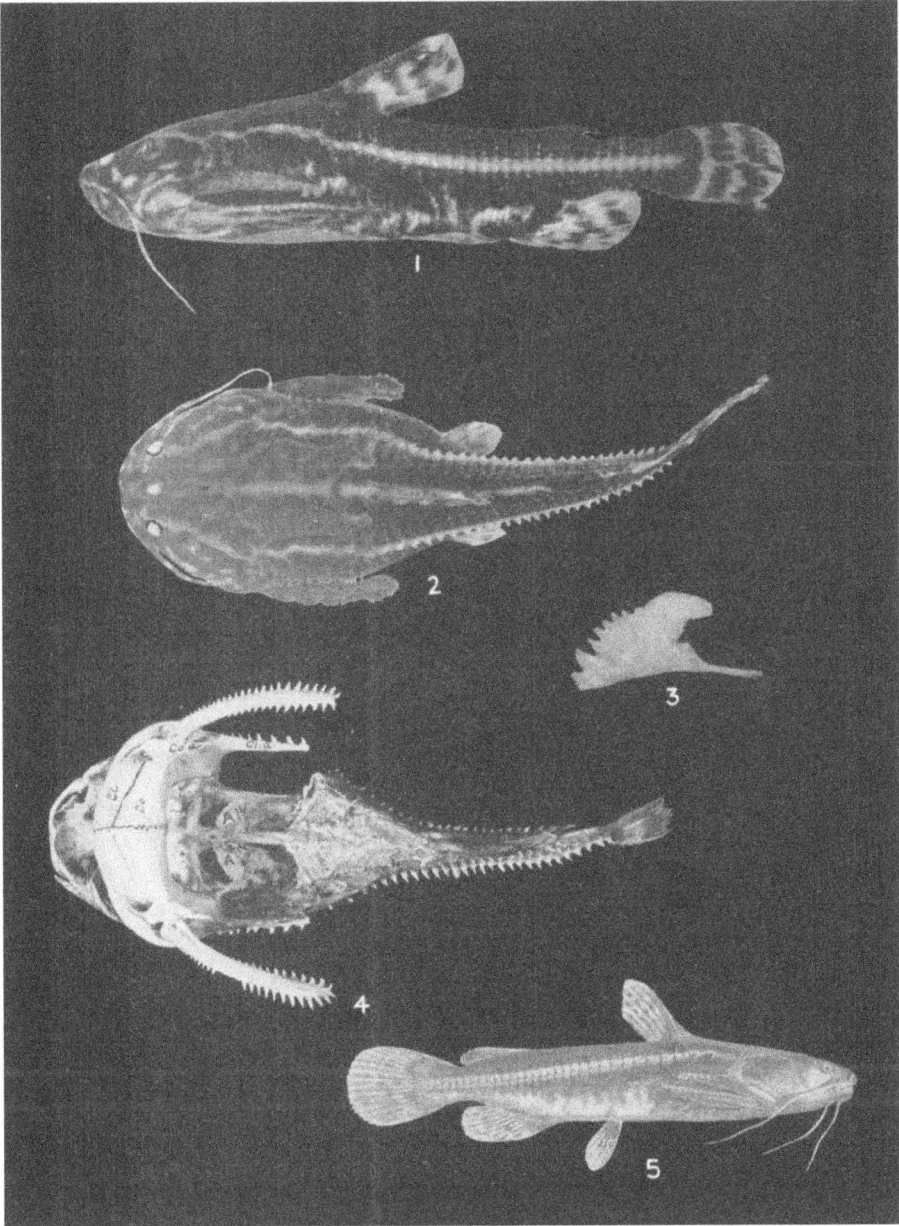

PLATE XI.

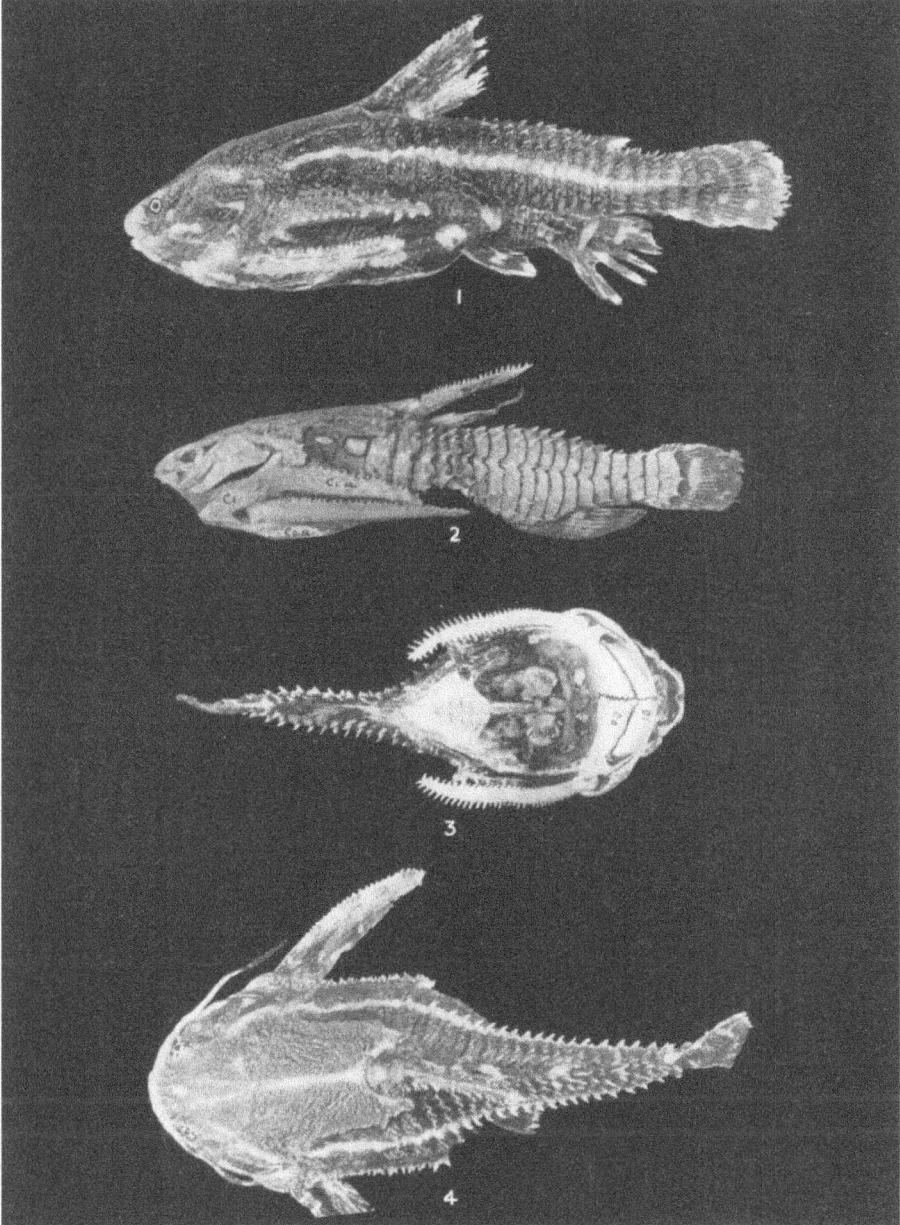

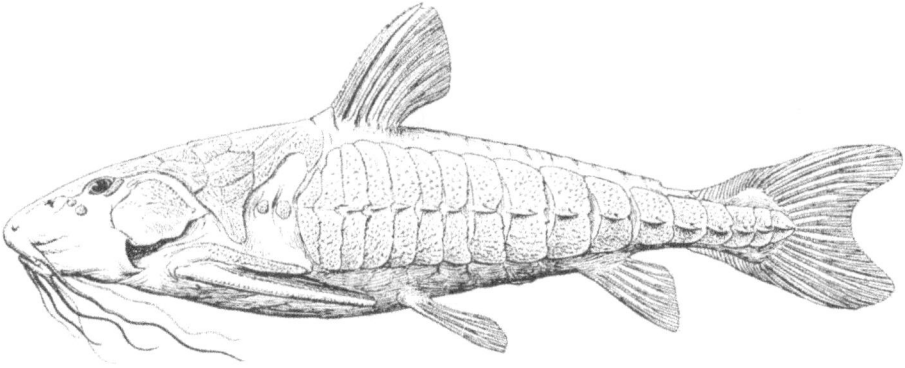

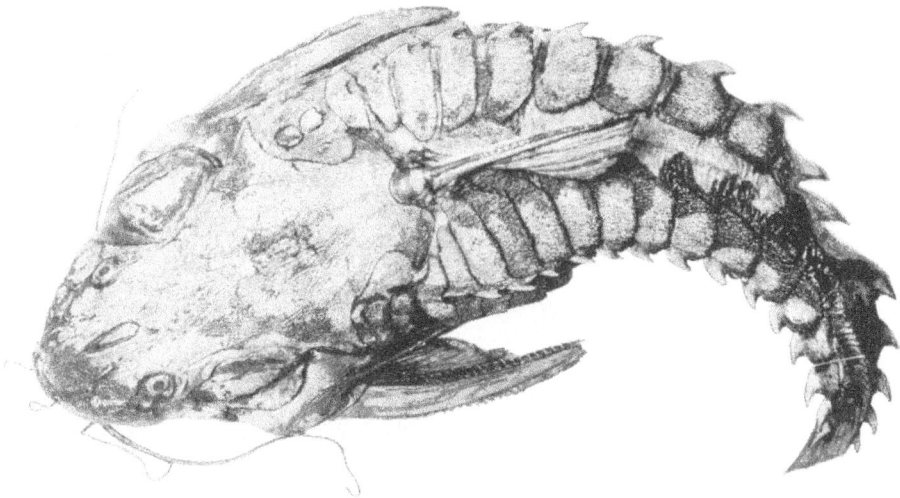

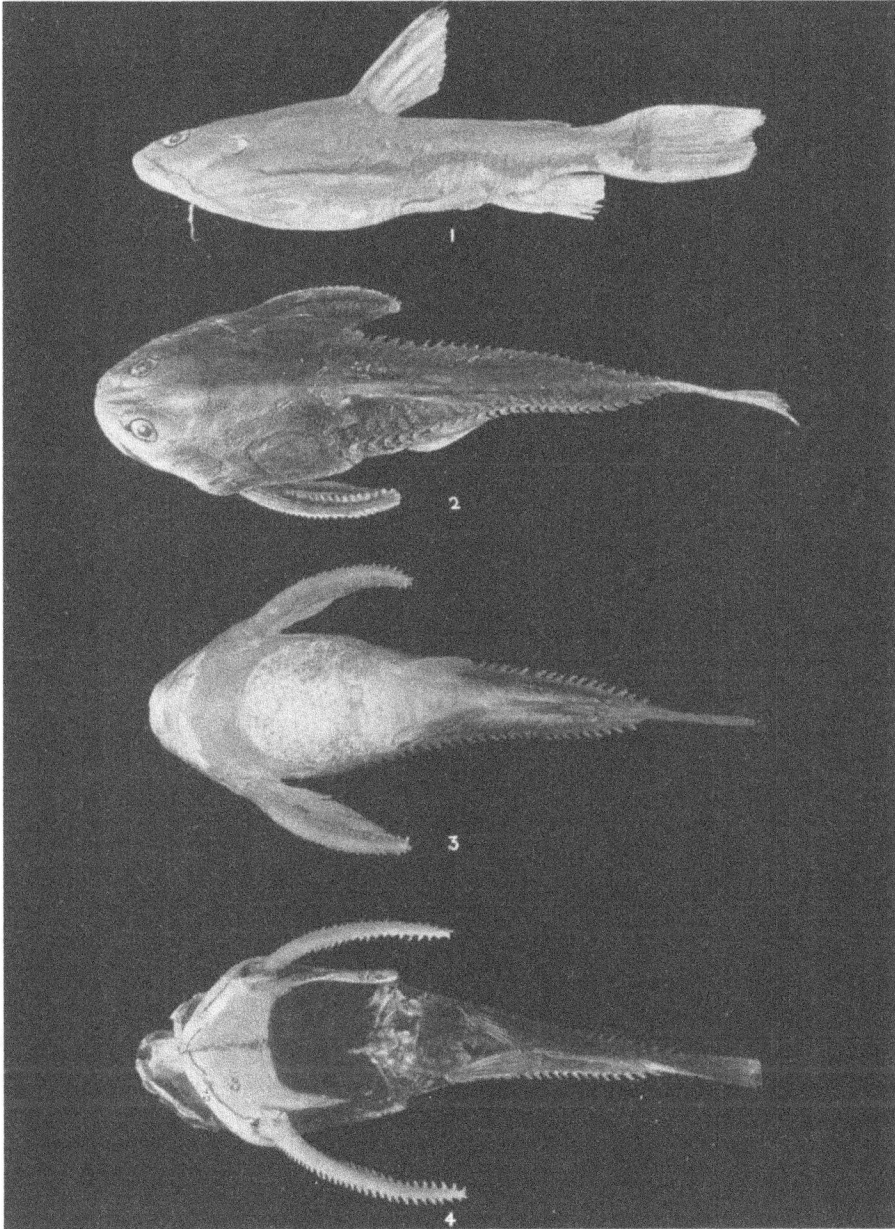

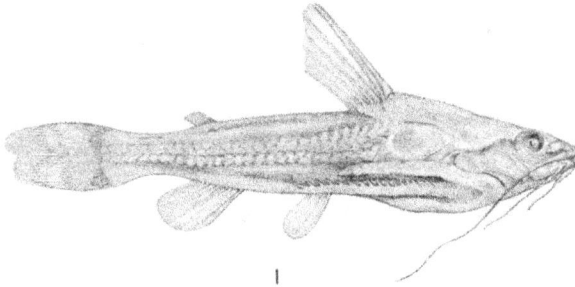

1

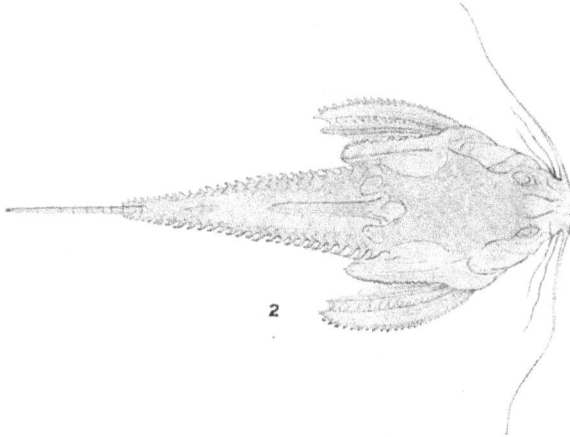

2

3

PLATE XV.

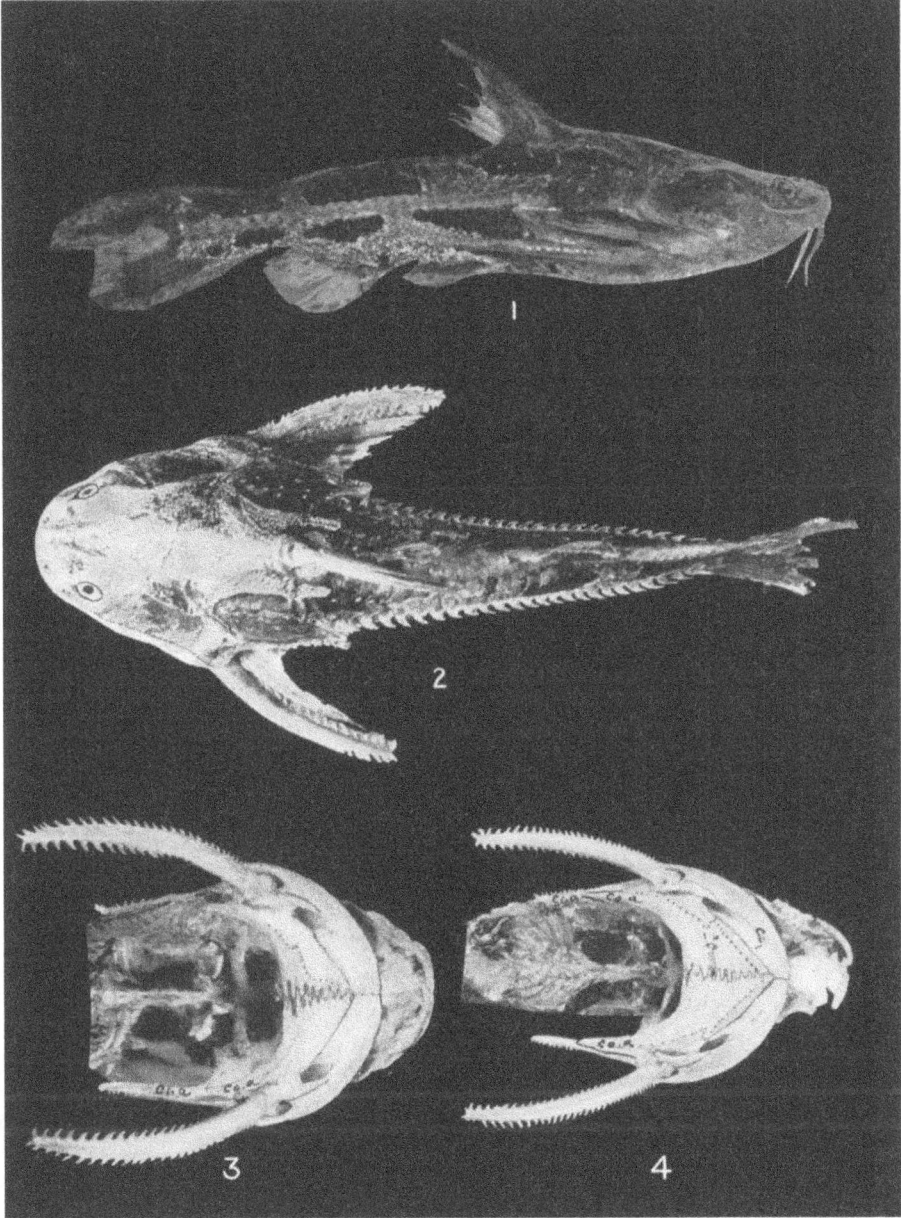

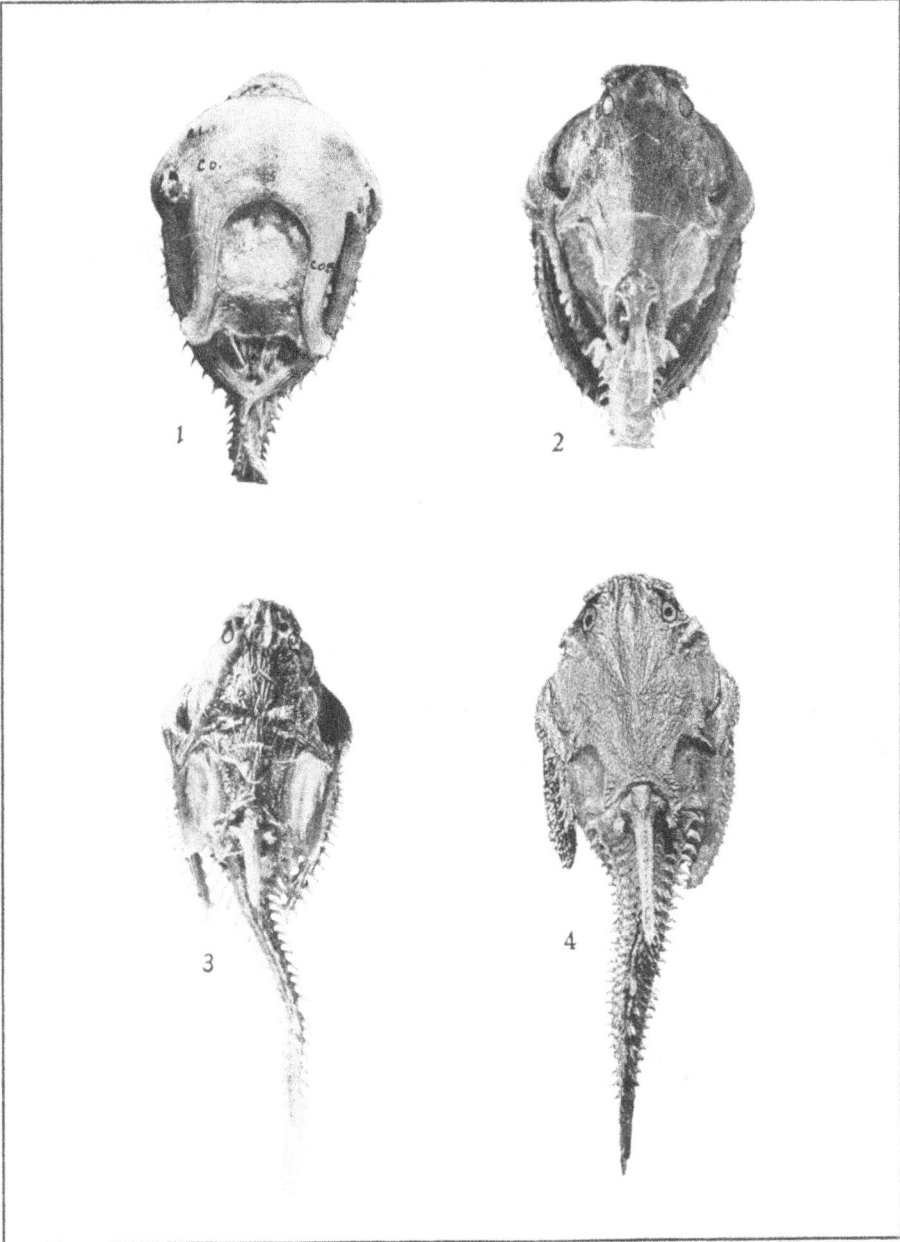

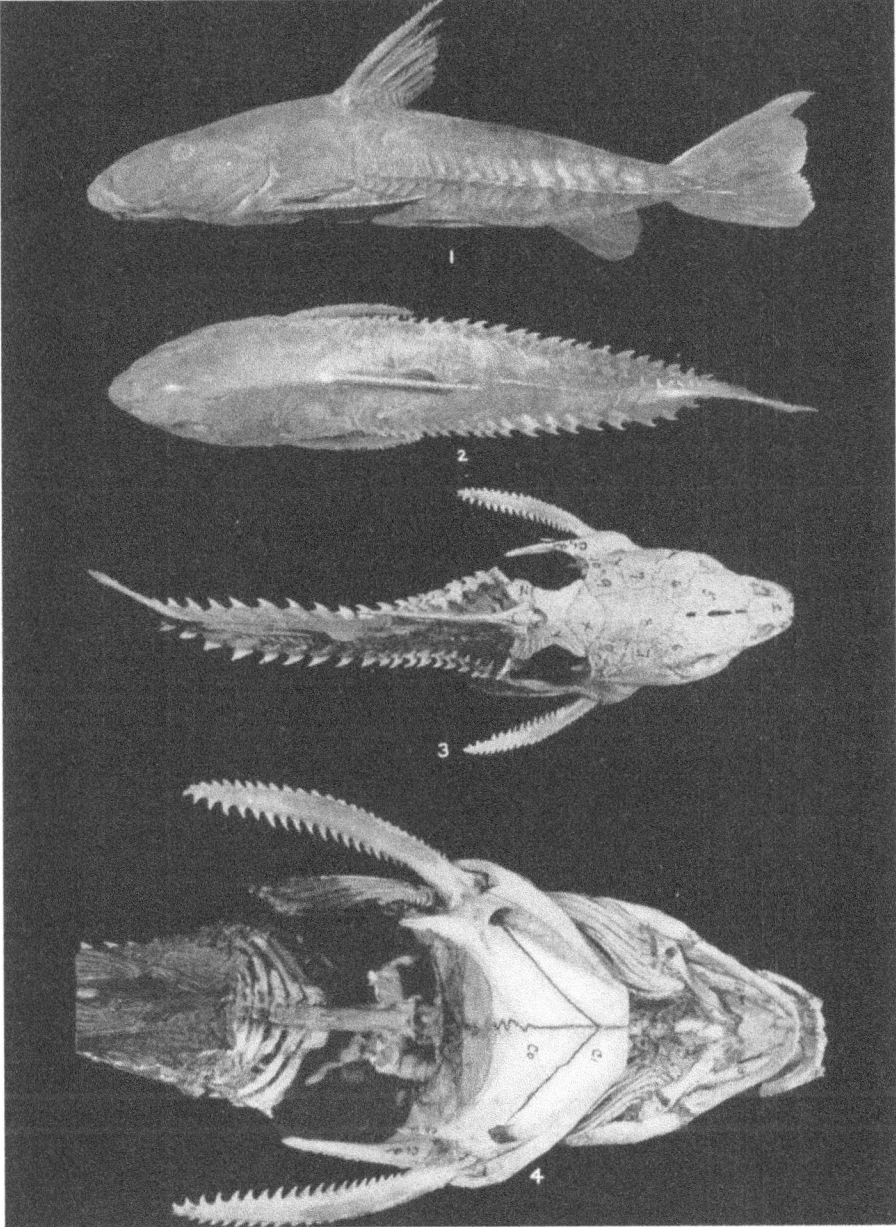

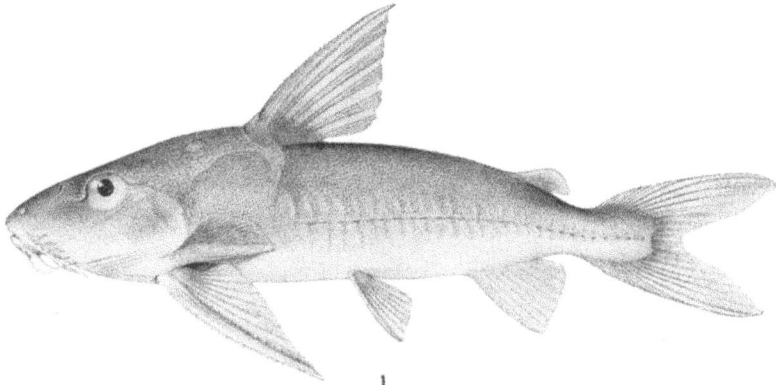

PLATE XX.

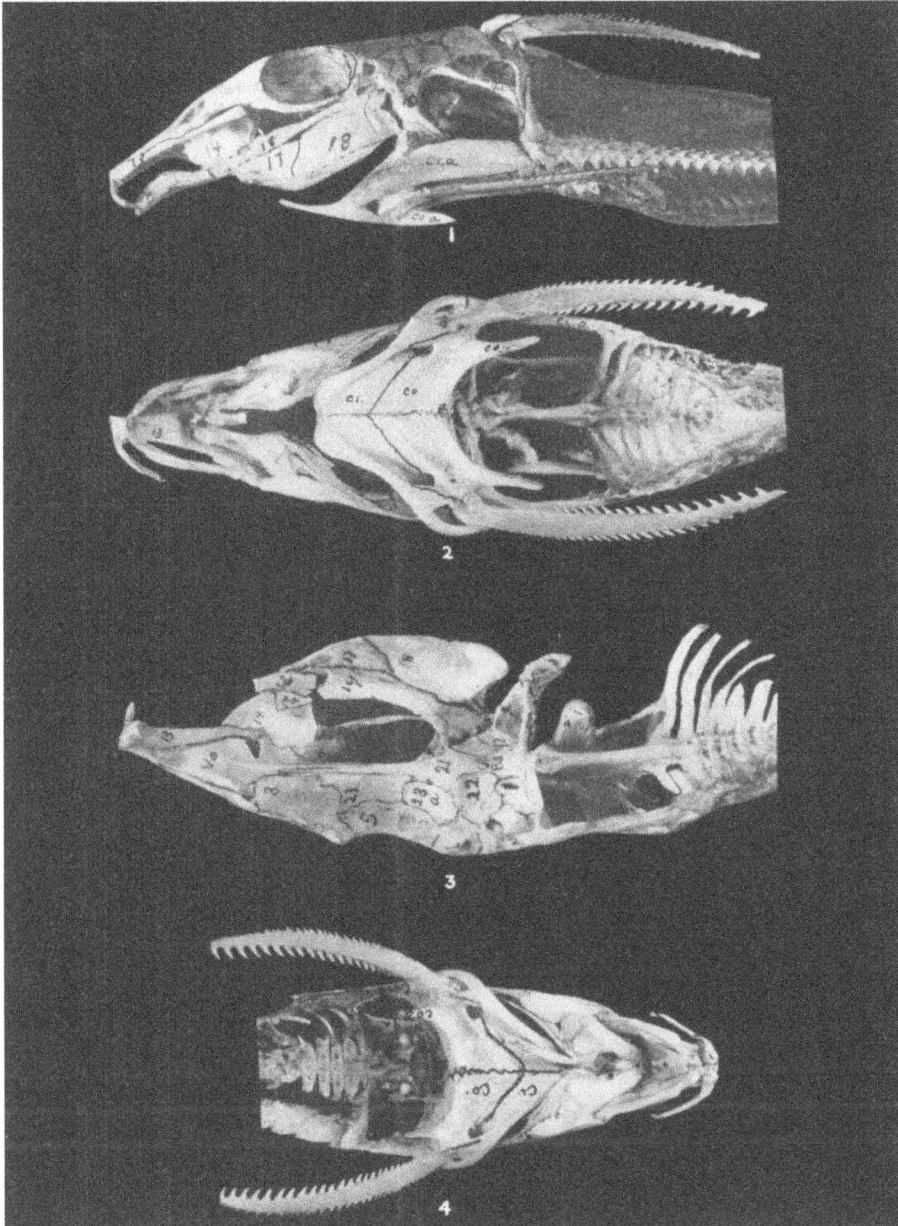

PLATE XXIII.

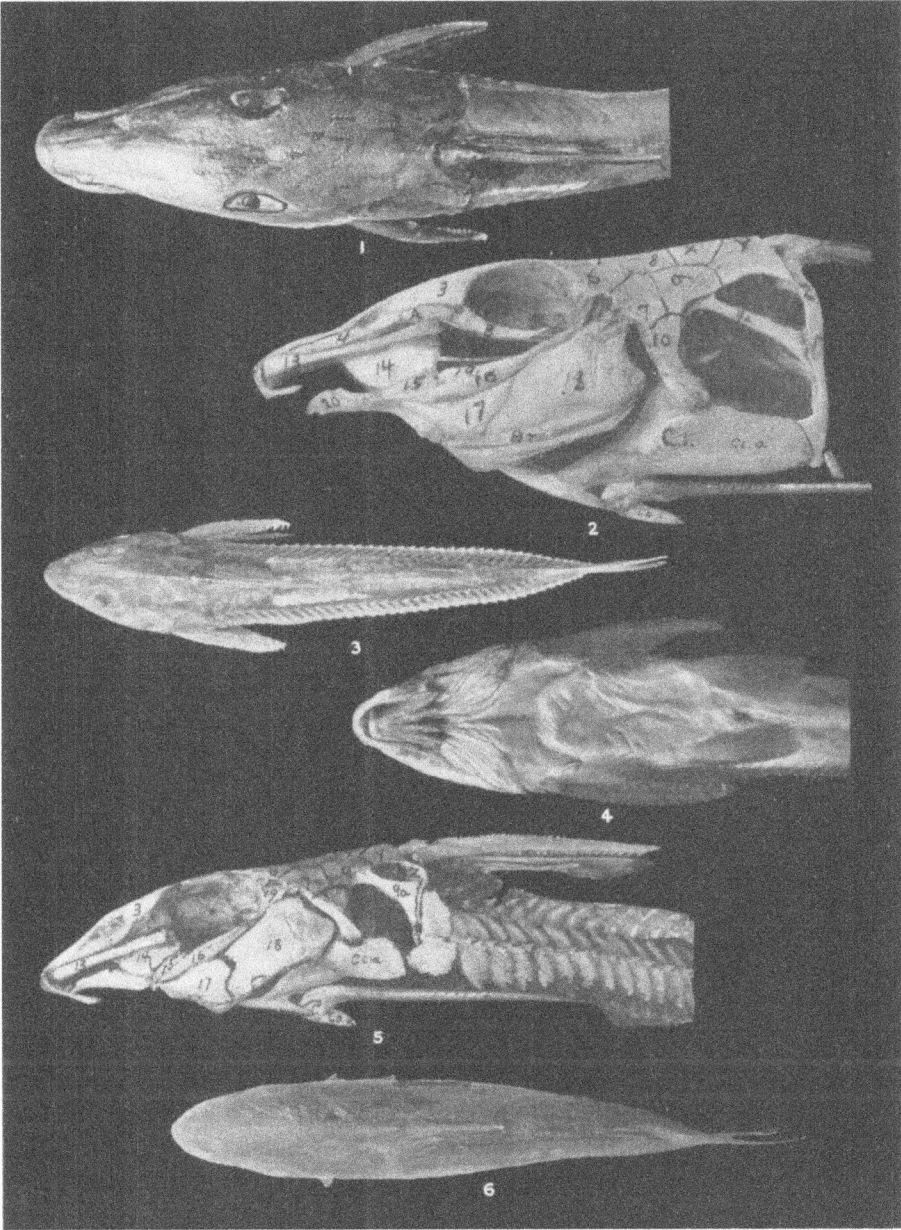

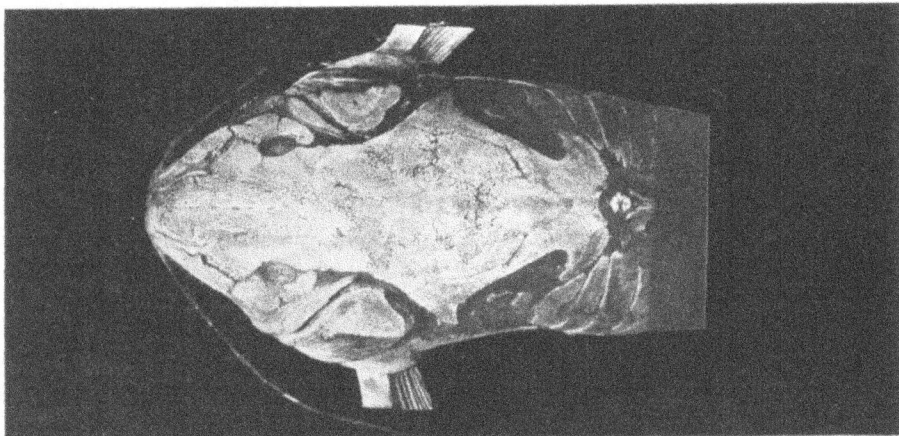

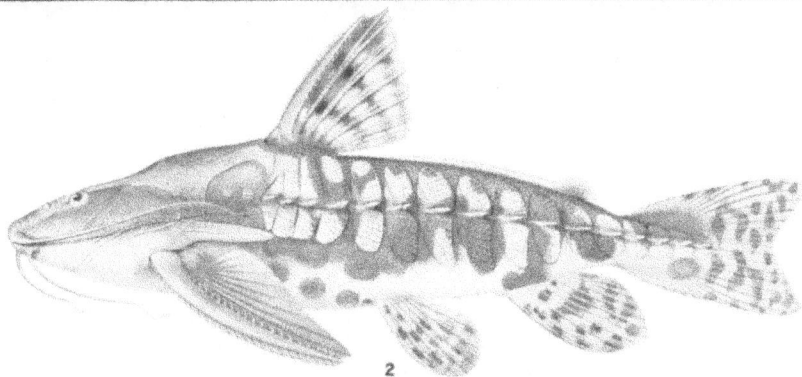

2

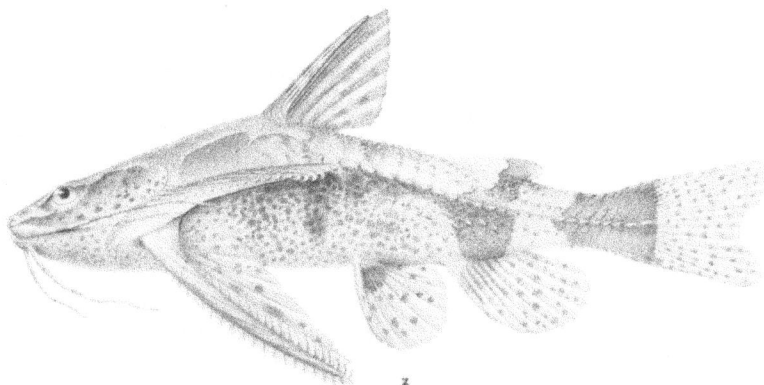

3

PLATE XXVI.

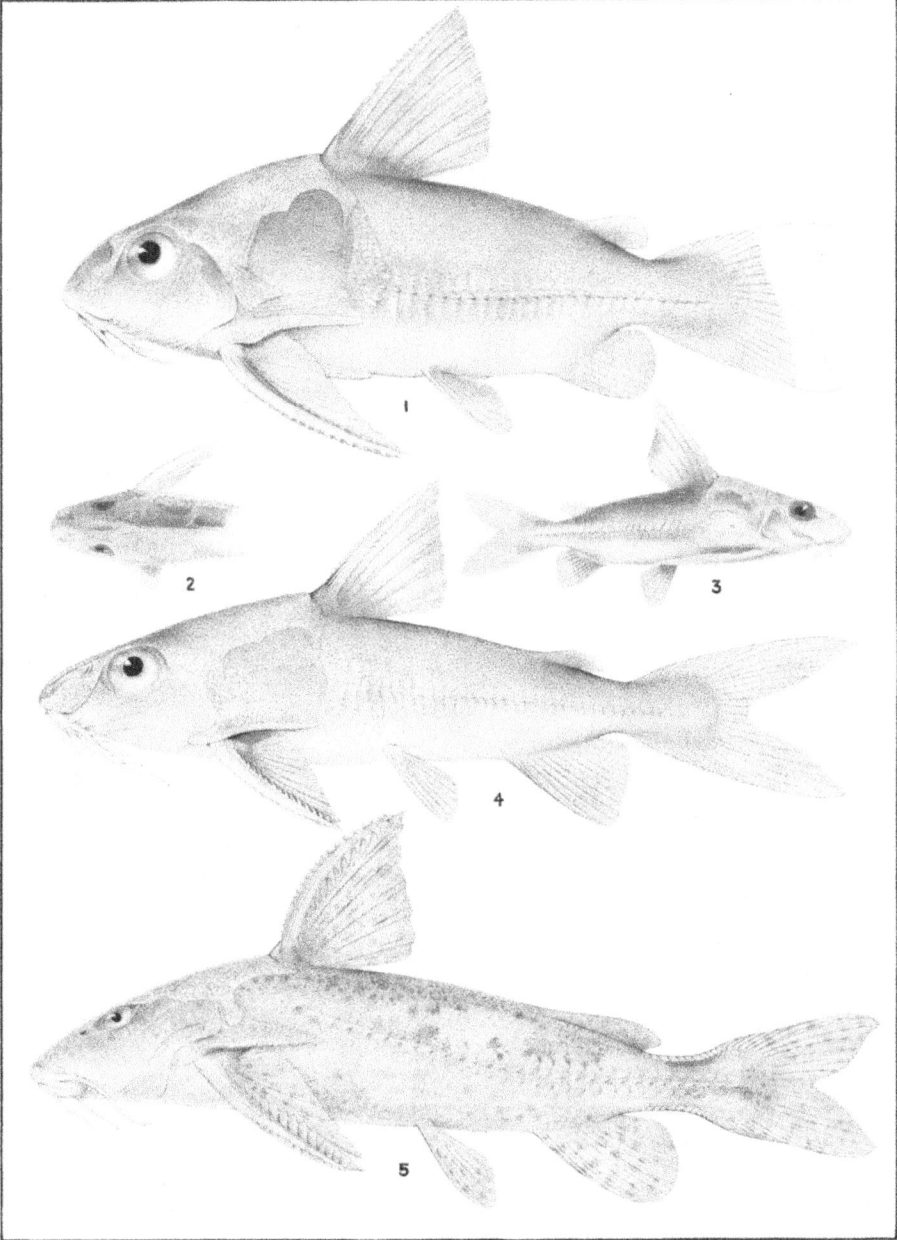

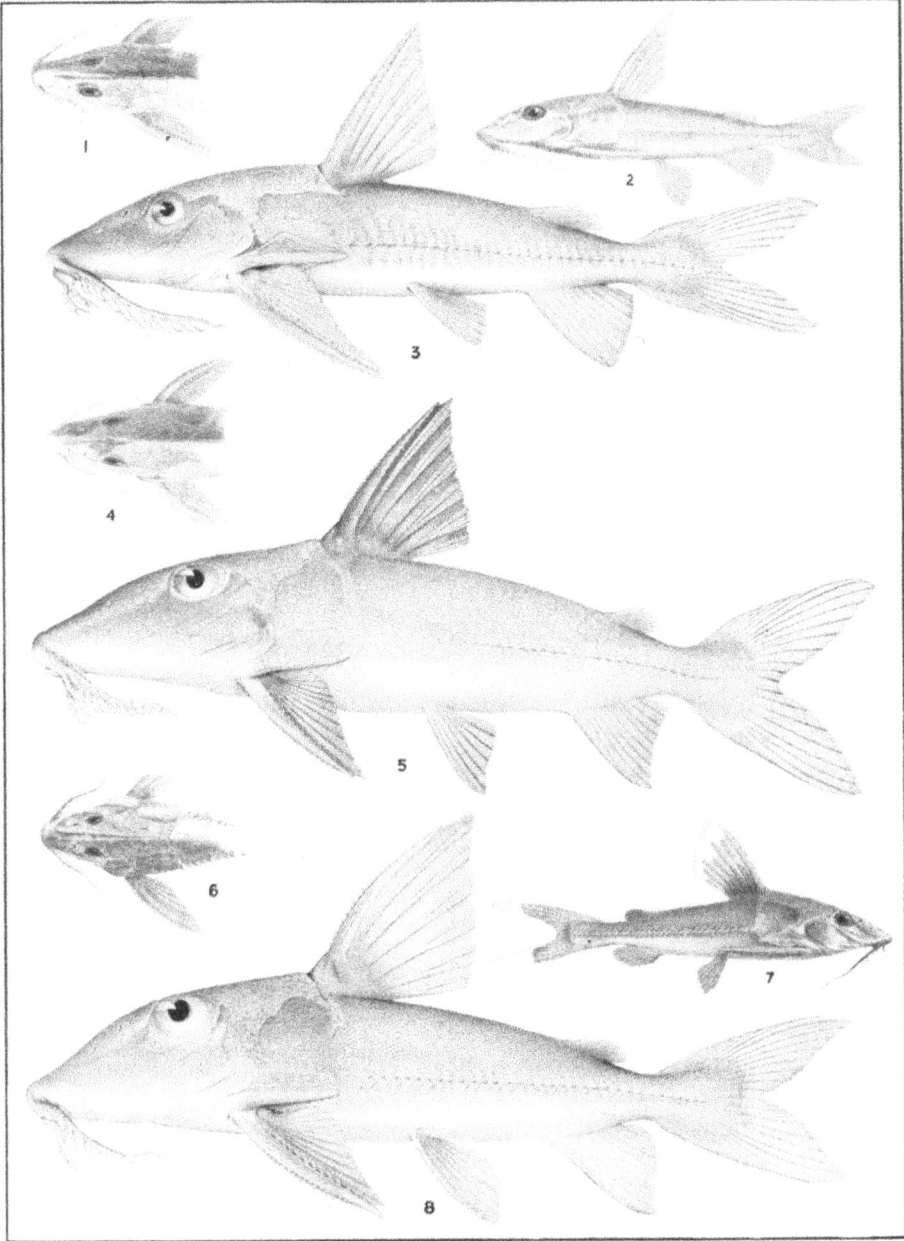

www.ingramcontent.com/pod-product-compliance
Lightning Source LLC
Chambersburg PA
CBHW081337190326
41458CB00018B/6032